Sustainable Building Performance

J. Vijayalaxmi
Editor

Sustainable Building Performance

Towards a Greener India

Springer

Editor
J. Vijayalaxmi
Department of Architecture
School of Planning and Architecture,
Vijayawada
Vijayawada, Andhra Pradesh, India

ISBN 978-981-96-2427-0 ISBN 978-981-96-2428-7 (eBook)
https://doi.org/10.1007/978-981-96-2428-7

This Springer imprint is published by the registered company Springer Nature Singapore Pte Ltd.
The registered company address is: 152 Beach Road, #21-01/04 Gateway East, Singapore 189721, Singapore

Dedicated to all my Students

Preface

This book highlights the evaluation and assessment of various aspects of sustainable buildings in order to achieve a greener future. India, at the 26th session of the United Nations Framework Convention on Climate Change (COP 26) in November, 2021, announced its target to achieve net zero by 2070. Being an emerging economy, the number of buildings required is expected to grow exponentially. Hence, there is a need for a pragmatic and realistic path to reach the committed goals of reaching net zero. In order to participate in a global transition to clean, low-carbon energy systems which is also the Sustainable Development Goal (SDG11) and Climate action (SDG 13), it is important to research on cooling strategies for a tropical country like India. The India Cooling Action Plan clearly acknowledges that there is a deep connect between SDGs (especially SDG 13, pertaining to Climate Action) and Cooling. Passive cooling strategies need a micro level understanding of the impact of building materials on the thermal performance of buildings. SDG 13.1 deals with resilience and adaptive capacity to climate-related disasters. India has had a rich history of climate responsive vernacular architecture. Many lessons can be learnt from the way buildings were designed to be thermally comfortable. This book looks at how biophilic architecture can be dovetailed in building codes and thus respond to Sustainable Development Goals to create sustainable and resilient communities paving way towards achieving the goals of SDG 11. The book presents chapters of architectural research based on real time data obtained through field studies. The research papers presented in this book is a result of systematic research conducted after identification of a research gap through the latest review of literature and presents new knowledge which can help in making performance-based design decisions.

The book will serve as a useful guide to any Postgraduate Scholar or Research scholar working in the domain of Energy Efficiency and Sustainable Buildings. Besides, it serves to understand the role of architects in driving India to a Greener and sustainable future in terms of built environment.

I express my sincere thanks to Dr. Loyola D'Silva, publishing editor at Springer, Ms. Niraja Deshmukh and Ms. Kavitha Palanisamy, Production Editor of the book for their great support during the preparation of the final book.

Dr. J. Vijayalaxmi
Professor, Department of Architecture
School of Planning and Architecture,
Vijayawada
An Institute of National Importance
Under the MoE
Government of India
Vijayawada, India

Contents

Sustainable Buildings for a Net-Zero India—Developments, Needs, and Challenges

J. Vijayalaxmi

Abstract Cooling energy demand would significantly increase if the earth were too warm by 1.5 to 2.0 °C, which is what is predicted. The world has taken initiatives to reduce this global warming. Simultaneously, India has also taken measures to reduce its GHG emissions, especially in the building sector. This is of significance because as a developing country, the demand for new buildings is on a steady rise. The sustainability issues taken by the world and also by India are discussed in a chronological sequence. Of much significance is the discussion, policies, and regulations pertaining to 'cooling'. India lies on the Tropics, much like many of the developing countries, and therefore, it is important to strategize passive 'cooling'. The sustainable development goal is another significant force to develop policies and strategies in India. The importance given to 'Cooling' in the sustainable development goals is also discussed. Passive strategies to bring outdoor and indoor thermal cooling are of utmost importance in the Indian context, as it can significantly reduce the load on air conditioners and thus reduce global warming. Field-based studies of thermal performance by use of passive strategies such as the use of vernacular building materials, low-carbon sequestered materials, biophilic architecture, vernacular buildings, assessment of CO_2 emissions, and reduction in embodied energy form the main crux of this book.

Keywords Global warming · Green House gas emissions · Embodied energy · Passive cooling · SGD · Vernacular construction

1 Global Scenario of Temperature Rise

Limiting a rise in the global mean temperature to 1.5 °C is becoming increasingly unachievable, as evidenced by an increasing number of studies. Heat exposure and cooling energy demand would significantly increase if the earth were too warm by 1.5 to 2.0 °C. A study by Miranda et al. (2023) indicates that the greatest rise in cooling

J. Vijayalaxmi (✉)
School of Planning and Architecture, Vijayawada, India
e-mail: vijayalaxmij2@gmail.com; vijayalaxmij@spav.ac.in

J. Vijayalaxmi (ed.), *Sustainable Building Performance*,
https://doi.org/10.1007/978-981-96-2428-7_1

demand will occur in areas close to the equator, especially in the sub-Saharan region between 2011 and 2020, and the global surface temperature has already risen by 1.09 °C above preindustrial levels.

The overall difference in cooling demand between now and a planet that is 2.0 °C warmer would be larger than our analysis maps, necessitating a critical focus on a topic that has historically been overlooked in discussions about sustainability.

The tropics and subtropics are covered by a small band that runs nearly parallel to the equator, and this zone contains the majority of the world's requirement for space cooling (Fig. 1). That potential is more than ten times greater in the hottest regions of the world, like the Middle East, Africa, India, and the Caribbean, than in the temperate zones, which include most of North America, western Europe, southern Argentina, and Chile, according to cooling degree days (CDDs), which quantify the positive deviation of temperatures from a reference point in a specific location over a given period of time. Africa, Latin America, southern and eastern Asia, and the Middle East have the largest absolute increases in CDDs, with increases ranging from 15 to 40% (Fig. 2).

Urgent, localized adaptation is required because even a slight increase in global warming raises heat exposure and cooling demand. With major effects on cooling degree days, both global North and South countries need to strive toward the 1.5 °C reduction goal. Resilience to warmer temperatures must be given top priority in energy and climate policies, and the unequal increase in cooling demand must be addressed with sociotechnical solutions that allow for comprehensive, sustainable responses.

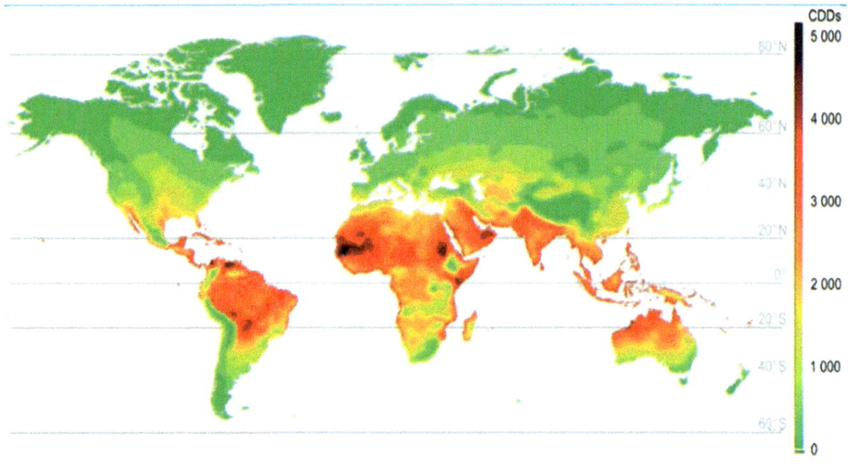

Fig. 1 Cooling degree days across the world, mean annual average (2007–2017)

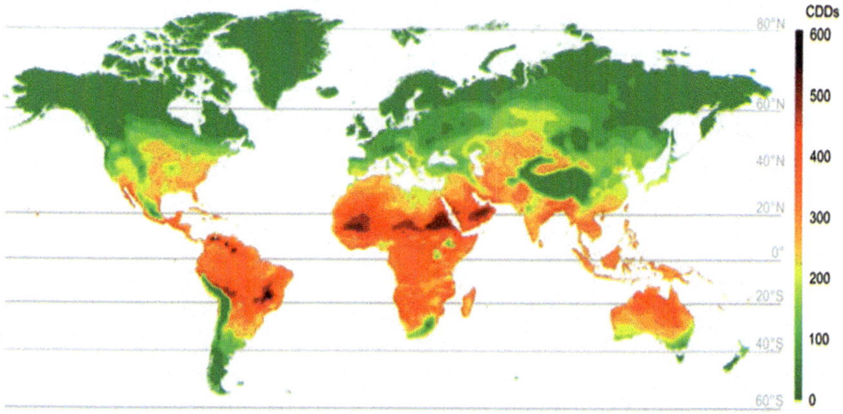

Fig. 2 Increase in cooling degree days in the baseline scenario relative to historical CDDs, 2016–2015. *Source* Ozone Cell (2023)

2 Global Sustainable Initiatives

The pursuit of sustainable architecture has evolved significantly over the last century. The roots of this movement can be traced back to the early twentieth century when architects began questioning the environmental impact of the buildings they created with scant regard to local climate and its vernacular architecture. The 1970s energy crisis served as a major turning point, raising awareness about the inefficiency of modern building designs (of the International Style of Architecture), and their contribution to energy consumption. During this time, architects and urban planners started exploring ways to reduce energy use, leading to the birth of the 'green building' concept. The focus was primarily on improving the energy efficiency of buildings through better insulation, the use of renewable energy sources, and integrating of passive design strategies.

2.1 1990—IPCC Report

The 1990 Intergovernmental Panel on Climate Change (IPCC) Report marked a critical turning point in the global recognition of climate change as a pressing environmental concern. Established in 1988 by the United Nations and the World Meteorological Organization, the IPCC's first assessment report underscored the mounting evidence that human activities, particularly the combustion of fossil fuels, were significantly increasing the concentration of greenhouse gases in the atmosphere (IPCC, 1990). It emphasized the need for a coordinated global response to mitigate the impacts of global warming, calling for a reduction in emissions and the promotion of energy-efficient technologies. By highlighting the dangers of continued greenhouse gas emissions, the IPCC report reinforced the urgency of adopting sustainable practices in diverse sectors, such as energy, agriculture, and urban development. The 1990 assessment was instrumental in fostering an international consensus on the importance of climate action and provided a roadmap for future climate agreements and sustainability efforts.

2.2 1992—UNFCCC

Formed during the Earth Summit in Rio de Janeiro, the Convention represented the first international treaty aimed at combating climate change through coordinated global action (UNFCCC, retrieved 2024). Its primary objective was to stabilize greenhouse gas concentrations in the atmosphere at a level that would prevent dangerous interference with the climate system, a goal that emphasized both mitigation and adaptation strategies. While the treaty itself did not impose binding emission reduction targets, it laid the foundation for more stringent commitments in the years that followed, particularly through the Kyoto Protocol in 1997 and the Paris Agreement in 2015. The Convention also introduced the principle of 'common but differentiated responsibilities', acknowledging that developed nations, due to their historical contributions to global emissions, should take the lead in reducing emissions and provide financial and technological support to developing countries.

2.3 1997—Kyoto Protocol

The Kyoto Protocol, adopted on December 11, 1997, marked a significant milestone in international efforts to combat climate change. After a complex ratification process, the Protocol entered into force on February 16, 2005. To date, there are 192 Parties to the Kyoto Protocol, including both countries and regional economic integration organizations. The Protocol operationalizes the United Nations Framework Convention on Climate Change (UNFCCC) by obligating industrialized nations and economies in transition to limit and reduce greenhouse gas (GHG) emissions according to individually agreed-upon targets. The Kyoto Protocol establishes legally binding commitments for emissions reductions. Central to the Protocol's structure is the principle of 'common but differentiated responsibilities and respective capabilities', These targets collectively aim for an average reduction of 5% in greenhouse gas emissions compared to 1990 levels during the first commitment period (2008–2012).

2.4 2015—The Paris Agreement

The Paris Agreement, adopted on December 12, 2015, at the UN Climate Change Conference (COP21) in Paris, is a legally binding international treaty aimed at addressing the growing threat of climate change. Entering into force on November 4, 2016, the Agreement set forth the ambitious goal of limiting the global temperature increase to 'well below 2 °C above pre-industrial levels', with an aspiration to limit the rise to 1.5 °C. This target is critical, as research by the UN's Intergovernmental Panel on Climate Change (IPCC) indicates that surpassing the 1.5 °C threshold could lead to far more severe impacts, including extreme heat waves, droughts, and floods. To meet this target, global greenhouse gas emissions must peak no later than 2025 and decline by 43% by 2030. Under this system, countries are required to submit nationally determined contributions (NDCs), which outline the actions they will take to reduce emissions and adapt to climate change. Each successive NDC must reflect greater ambition compared to the previous submission. At the COP27 conference, the decision was made to encourage Parties to revisit and strengthen their 2030 targets to align with the 1.5 °C goal by the end of 2023, taking into account national circumstances. In addition to NDCs, the Paris Agreement invites countries to develop long-term low greenhouse gas emission development strategies (LT-LEDS). These strategies, while not mandatory, offer a broader, long-term vision that guides countries' development plans and aligns them with their climate goals. A crucial element of the Paris Agreement is its emphasis on climate finance. Developed nations are urged to lead in providing financial support to developing countries, with additional contributions encouraged from other Parties. This financial assistance is essential not only for mitigation efforts, such as large-scale emissions reductions, but also for adaptation measures, which help vulnerable countries cope with the impacts of climate change. Furthermore, the Paris Agreement highlights the importance of technology development and transfer to enhance resilience and reduce emissions. A dedicated Technology Mechanism has been established to accelerate the development and transfer of critical technologies. Developed nations are called upon to support capacity-building actions in these countries, helping them meet their climate goals. To ensure transparency and accountability, the Paris Agreement also established an enhanced transparency framework (ETF), under which countries will report on their climate actions starting in 2024. This framework will allow for the assessment of global progress toward the long-term climate objectives, with recommendations made for further action based on the findings of the Global Stocktake.

2.5 Sustainability and Green Building Initiatives

One of the most significant milestones for buildings came in 1993 with the creation of the U.S. Green Building Council (USGBC) and the introduction of the Leadership in Energy and Environmental Design (LEED) Green building certification system.

LEED quickly became a benchmark for sustainable building practices, promoting a holistic approach to architecture that incorporates energy efficiency, water conservation, sustainable materials, and indoor air quality. By the late 1990s, architects around the world were increasingly adopting these green building principles, and the concept of the 'Net-Zero' building—structures that produce as much energy as they consume, began to gain traction. The push for sustainable architecture reached new heights in 2015 with the Paris Agreement, where nations collectively pledged to limit global warming to below 2 °C, further solidifying the role of buildings in climate action. Then came a unified call for regenerative buildings—buildings that produce more electricity than they consume.

3 India—Sustainability Initiatives

In the light of the global initiative to reduce GHG, India has taken several steps to reduce its GHG and contribute to reducing global warming. The Indira Paryavaran Bhawan in New Delhi, completed in 2011, became one of the first government buildings to achieve a LEED Platinum certification, showcasing the potential for large-scale sustainable design in India's urban centers. Other projects like the CII-Sohrabji Godrej Green Business Centre in Hyderabad demonstrated the country's growing commitment to sustainable architecture, with their emphasis on energy-efficient systems, water conservation, and sustainable materials and their commitment to capacity building of Green Buildings. In recent years, the focus on energy-efficient buildings has intensified, with projects like the Sakthi Sustainable Energy Building in Coimbatore leading the way in integrating solar power, rainwater harvesting, and passive design techniques. These developments not only reflect a shift toward sustainability in urban centers but also highlight the importance of designing for India's diverse climate zones and resource availability. Other initiatives over a period of time are as follows.

3.1 2001—Energy Conservation Act, IGBC Formation

In 2001, India took a major step forward with the Energy Conservation Act, a landmark piece of legislation aimed at reducing energy consumption across sectors. This year also saw the formation of the Indian Green Building Council (IGBC), which played a pivotal role in promoting sustainable building practices in the country (IGBC, 2020).

3.2 2002—Creation of Bureau of Energy Efficiency

The following year, in 2002, the Bureau of Energy Efficiency (BEE) was established, tasked with driving energy efficiency efforts across industries, businesses, and households. The mission of the Bureau of Energy Efficiency is to assist in developing policies and strategies with a thrust on self-regulation and market principles, within the overall framework of the Energy Conservation Act, 2001 with the primary objective of reducing energy intensity of the Indian economy. The major regulatory functions of BEE include

- Develop minimum energy performance standards for equipment and appliances under Standards and Labelling
- Develop minimum energy performance standards for Commercial Buildings
- Develop Energy Consumption Norms for Designated Consumers
- Certify energy managers and energy auditors.

3.3 2005—GRIHA Launched, NBC Revised

In 2005, two significant milestones occurred: the launch of the Green Rating for Integrated Habitat Assessment (GRIHA), a performance-based rating system for green buildings, and the revision of the National Building Code (NBC) to incorporate more sustainable construction practices. GRIHA has been adopted by the Ministry of New and Renewable Energy to 'rate' a building on the degree of its 'greenness'. The agenda is to benefit the community at large with the improvement in the environment by reducing GHG (greenhouse gas) emissions, reducing energy consumption, and the stress on natural resources. These developments signaled India's growing commitment to green building standards.

3.4 2007—11th Plan, ECBC Launched, LEED-India Launched

The 11th Five-Year Plan (2007–2012) marked a turning point, with the launch of the Energy Conservation Building Code (ECBC), which set minimum energy performance standards for commercial buildings. It provides existing as well as futuristic advancements in building technology to further reduce building energy consumption and promote low-carbon growth Additionally, LEED-India was introduced, adapting the international Leadership in Energy and Environmental Design (LEED) certification to local conditions, further promoting sustainable architecture in India.

3.5 2008—National Action Plan on Climate Change (NAPCC), ECBC Revised

In 2008, India launched the National Action Plan on Climate Change, a comprehensive framework to address the country's climate-related challenges. The NAPCC outlined eight key missions, each focused on different areas of climate action, from renewable energy to energy efficiency, with the goal of fostering sustainable development while reducing India's carbon footprint. During this year, the Energy Conservation Building Code (ECBC) was also revised to ensure that newer commercial buildings adhered to stricter energy efficiency norms.

3.6 2009—BEE Energy Star Rating Launched for Commercial Buildings

The year 2009 saw the introduction of the BEE Energy Star Rating for commercial buildings, a voluntary program designed to help building owners and managers assess their energy efficiency performance. This year also marked the establishment of the Global Environment Facility (GEF-5) project, aimed at fostering energy efficiency and sustainable development.

3.7 2010—NAPCC Mission on Sustainable Habitat, GEF-5, EC Act Amended

In 2010, a crucial addition to the NAPCC came with the launch of the Sustainable Habitat Mission. It emphasized the importance of creating energy-efficient buildings, sustainable urban transport systems, and reducing energy consumption in residential and commercial sectors. It encouraged the adoption of green building practices, including energy-efficient construction techniques, to reduce the overall carbon footprint of urban spaces. This year also saw the amendment of the Energy Conservation Act, which further strengthened the legal framework for energy efficiency. Additionally, the NAPCC Mission on Enhanced Energy Efficiency was initiated, providing further momentum to energy efficiency efforts.

3.8 2011—NAPCC Mission on Enhanced Energy Efficiency, Rajasthan and Odisha Notify ECBC, UNDP-GEF-BEE Project

In 2011, two Indian states, Rajasthan and Odisha, formally notified the Energy Conservation Building Code (ECBC), making it mandatory for commercial buildings in these regions to comply with energy efficiency standards. The UNDP-GEF-BEE project was also launched, further supporting India's push for sustainable energy practices.

3.9 2012—12th Plan, UNDP-GEF-BEE PMU Established

By 2012, the 12th Five-Year Plan came into effect, which included a continued focus on sustainability and energy efficiency. Additionally, the UNDP-GEF-BEE Project Management Unit (PMU) was established to oversee the implementation of energy-efficient initiatives across India, strengthening the country's commitment to meeting global sustainability targets.

3.10 Eco Niwas Samhita

In 2018, the Eco Niwas Samhita (ENS) was introduced by the Ministry of Power under the Bureau of Energy Efficiency (BEE). The initiative aimed at improving the energy performance of residential buildings, aligning with India's commitment to energy efficiency and climate action. Eco Niwas Samhita provides guidelines and requirements for energy-efficient design and construction, focusing on aspects such as thermal comfort, lighting, and energy-efficient building materials. This initiative primarily targets new residential buildings, with an emphasis on reducing energy consumption and greenhouse gas emissions. ENS outlines key measures, including the use of efficient insulation, natural lighting, and the integration of renewable energy sources, to ensure the built environment contributes to India's broader sustainability goals.

3.11 Green Credits Program (2017)

Launched in 2017, the Green Credits Program is an initiative by the Indian Green Building Council (IGBC) aimed at encouraging sustainability in the construction sector. The program provides a framework for recognizing buildings that go beyond traditional green certifications like LEED. The Green Credits Program allows building owners and developers to earn credits for implementing advanced sustainability measures, including energy conservation, water efficiency, use of sustainable materials, and waste management. The goal is to incentivize innovation in sustainable construction practices and help India achieve its climate goals, particularly in the context of rapidly growing urban populations and infrastructure needs.

3.12 Retrofitting of Existing Buildings in India

As part of India's broader commitment to energy efficiency, the retrofitting of existing buildings has gained significant attention in recent years. With a large portion of India's built environment consisting of older structures, retrofitting efforts are critical to reducing energy consumption and enhancing sustainability. In 2016, the Bureau of Energy Efficiency (BEE) launched initiatives aimed at improving the energy performance of existing buildings. The retrofitting process involves upgrading buildings with energy-efficient systems, such as better insulation, energy-efficient lighting, and the integration of renewable energy sources like solar panels. These efforts are particularly important in the context of India's 2030 climate targets, as existing buildings are a major source of energy use and carbon emissions. Additionally, retrofitting is a cost-effective way to improve the sustainability of the building stock without the need for complete redevelopment. The government has also promoted the concept of 'Energy Service Companies (ESCOs)', which provide expertise in retrofitting and help manage the financing of energy efficiency upgrades.

4 India's Net-Zero Challenge

India faces a unique challenge in its pursuit of Net-Zero emissions. While the country is committed to reducing its carbon footprint, it also has to account for the rapid urbanization and economic growth that are essential for improving the quality of life of its burgeoning population. With coal accounting for more than 70% of India's electricity generation, the path to a low-carbon future seems daunting. The country's energy demands are rising rapidly, driven by industrialization, urban growth, and an expanding middle class. Yet, India's commitment to Net-Zero emissions by 2070—announced at the 2021 COP26 summit—reflects the government's recognition of the need for action (Ahluwalia and Patel, 2021). India's challenge lies not only just in its reliance on coal but also in the need to develop its infrastructure at a pace that meets the demands of an expanding population. India must balance its aspiration to become a global economic powerhouse with the necessity to minimize environmental degradation. As a developing nation, it is also crucial that India secures financial and technological support from developed countries to transition toward renewable energy and sustainable practices. Global forums have increasingly recognized India's unique position in the climate crisis, acknowledging the need for a just energy transition that doesn't stifle growth. India's transition to Net-Zero will require a multifaceted approach, including the promotion of renewable energy sources, energy-efficient buildings, and sustainable urban planning practices.

5 Challenges in Passive Cooling for India's Built Environment

A significant challenge India faces in its path to sustainable architecture is the difficulty in implementing passive cooling techniques in the built environment, particularly in densely populated urban areas. Traditional Indian architecture often relied on passive cooling through natural ventilation, the strategic use of courtyards, and the placement of windows and doors to promote airflow. However, these techniques have become less effective in today's rapidly urbanizing cities, where space is limited, and the urban heat island effect exacerbates the need for cooling (Panchabikesan et al., 2017). In cities like Delhi, Bengaluru, and Mumbai, where dense populations and large-scale construction have replaced traditional buildings, the application of passive cooling methods is becoming increasingly difficult. The prevalence of concrete and glass, materials that trap heat, further exacerbates the issue. Additionally, the socioeconomic disparities in India make it challenging to implement energy-efficient cooling systems across all sectors of society (Iqbal et al., 2021). To address these challenges, India must explore innovative cooling solutions, such as hybrid systems that combine passive and active cooling methods. Moreover, it is crucial to integrate passive cooling techniques into building codes and urban planning strategies, ensuring that new developments prioritize climate-responsive design.

6 Sustainable Development Goals in Fulfilling the Passive Cooling Initiatives

Adopted unanimously by all United Nations Member States in 2015, the 2030 Agenda for Sustainable Development provides a comprehensive framework for achieving a sustainable future. The 17 Sustainable Development Goals (SDGs) at its core represent a global commitment to eradicate poverty, reduce inequality, and protect the planet. These 169 targets, underpinned by the principle of 'leaving no one behind,' provide a detailed roadmap for achieving the Sustainable Development Goals. They emphasize the importance of addressing poverty, inequality, and environmental degradation simultaneously. (UN DESA, 2024).

The SDG calls upon countries of all incomes and status to improve the lives of their citizens. The SDGs have a universal scope. India's commitment to the 2030 Agenda has been affirmed by the Hon'ble Prime Minister of India in the UN Sustainable

Development Summit 2015, stating that much of India's development agenda mirrors that in the SDGs (Niti Aayog, 2024).

7 SDG and Cooling

Tropical countries like India need to primarily focus on cooling of buildings in order to meet the goals to reduce GHG. This section discusses the importance given to 'cooling' and its need in the SDG.

Goal 7 focuses on universal access, renewable energy and energy efficiency (in terms of GDP and primary energy). Goal 11 emphasizes on access to housing and transport, while reducing environmental impacts and disaster risks. Additionally, it also encourages heritage protection and resilience building. None of the SDGs specify passive cooling potential as a strategy for energy reduction and sustainable habitat. These techniques can lower electricity bills, create healthier living spaces, and reduce the strain on power grids (MoEFCC, 2024; UN DESA, 2024).

Heat stress, exacerbated by climate change, is a leading cause of weather-related deaths and worsens various health conditions, with heatstroke posing a significant fatality risk. Populations in tropical regions and low-quality housing, especially the urban and rural poor, are increasingly vulnerable to extreme heat, facing greater exposure due to lack of affordability to mechanical cooling (Gamero-Salinas et al., 2021). This can lead to overheating of interior spaces as well which can affect the inhabitants' health poorly. Passive cooling reduces this risk further improving the resilience of the built environment to the ever-increasing global temperatures. About two to four billion people need domestic space cooling to prevent the deadly heat risks that three-quarters of humanity will face under the current climate and socioeconomic conditions (Khosla et al., 2021). There is a large gap in access to essential space cooling in India, South-East Asia, and sub-Saharan Africa. This number surpasses the energy poverty gap as defined by the sustainable development goals. In order to meet this demand, an increase of 14% in current global residential electricity consumption would be required. This cannot be met only by mechanical means and requires passive cooling and other innovative cooling solutions to achieve this goal with minimum environmental impact (Lessio et al., 2019).

In spite of the present exponential need for cooling, it is not spoken about in the sustainability debates. Furthermore, the terms cooling and thermal comfort are overlooked in the UN's 2030 Agenda for Sustainable Development beyond the energy aspects. Each of the SDG is directly/indirectly influenced by cooling, and it is important for it to be factored in for successful SDG outcomes. Socioeconomic trends, technological availability, environmental trends driven by global climate change and geopolitical trends directly affect cooling demand and technologies (Khosla et al., 2021).

India has faced difficulties in achieving the SDGs due to barriers to decision making such as lack of climatic information and coordination among and within various ministries (Dhanapal et al., 2023).

The India Cooling Action Plan (ICAP) was published in 2019. It aims at cutting down refrigeration demand by 25–30%, cooling demand by 20–25% and cooling energy requirements by 25% to 40% by the year 2037. One of ICAP's six subject areas, Space Cooling in Buildings, focuses on encouraging the use of sustainable cooling systems in both new construction and building retrofits, which also includes passive cooling techniques (MoEFCC, 2024). The India Cooling Action Plan focuses on energy efficiency of equipment (Dixit et al., 2023).

The global average per capita energy consumption in space cooling is 272 kWh. India averages on the lower end with just 69 KWh. This is due to the income disparity with only 8% households owning an AC. As heat waves become more common, this number is bound to shoot up to attain minimum comfort standards. The increased cooling load leads to overloading of the grid and increase in the energy consumption (Carboncopy, 2024). This directly affects all the other SDGs and will set it backward over the years. The current solutions focus on energy efficiency by improving performance of electric appliances and demand side management. The ICAP report also mentions passive cooling solutions which require changes in building regulations. This has been proposed to be integrated with the Pradhan Mantri Awas Yojana (PWAY) scheme which aims at providing affordable housing to urban and rural poor (MoEFCC, 2022). Some of the drawbacks of the ICAP are the lack of set benchmarks which can interfere with the country's energy budget while cooling still remains unaffordable for many.

8 SDG as a Tool to Tackle Global Warming

Climate change is fueling a surge in extreme heat events, causing catastrophic consequences for urban areas. Tragically, many deaths linked to heat exposure are overlooked or misattributed, delaying essential measures to address this growing crisis. In a historic move, 193 countries, including India, united with the UN's 2030 Agenda for Sustainable Development in 2015. This ambitious plan, outlined in the SDGs, aims to tackle global challenges across social, economic, and environmental spheres.

8.1 India's Progress in Meeting the Sustainable Development Goals

India's commitment to the 2030 Agenda for Sustainable Development is evident in the concerted efforts undertaken by NITI Aayog to localize the SDGs. By working closely with States and UTs, NITI Aayog has sought to institutionalize the SDGs, ensuring their integration into national and subnational development policies and programs.

This approach, characterized by collaborative federalism and a whole-of-society perspective, aims to foster a sustainable development paradigm across the country. The NITI Aayog SDG India Index is a groundbreaking tool that tracks the progress of India's states and UTs in their quest for a sustainable future. The progress is monitored as shown in Fig. 3. By fostering cooperation and competition among different levels of government, this index aims to accelerate India's journey toward achieving the SDGs. It can be clearly seen that SDG 13, which is climate action needs more impetus. India, being a tropical country, needs to lay emphasis on strategies that enhance cooling in order to meet up to the goals of SDG.

9 Passive Cooling Need in India

Building on the progress made with the millennium development goals (MDGs), the SDGs climate change is fueling a surge in extreme heat events, causing catastrophic consequences for urban areas. Tragically, many deaths linked to heat exposure are overlooked or misattributed, delaying essential measures to address this growing crisis. Hence, there was a need to develop a 'Cooling Plan' for a tropical warm country like India. The India Cooling Action Plan (ICAP, 2019), clearly acknowledges that there is a deep connect between SDGs (especially SDG 13, pertaining to Climate Action) and cooling.

9.1 India's Geographical Scenario

The seventh-largest nation in the world, India, offers a wide range of landscapes. The majority of India experiences a tropical climate, reflecting the country's geographical variety. The country's varied topography results in a range of temperature regimes, from hot and cold, aridity and little rainfall to extreme humidity and heavy rainfall, as well as continental and coastal (Fig. 4). The nation is also susceptible to floods, droughts, cyclones, landslides, avalanches, storms, and heat waves to varied degrees due to its distinct geoclimatic conditions.

9.2 Cooling Demand in India

Despite its rapid urbanization and rising per capita income, India's economy is still characterized by low penetration of air-conditioning and refrigeration equipment, which is likely to increase the need for cooling in the near future.

In order to achieve planned economic growth and social development, as well as to satisfy numerous sustainable development goals, cooling is now acknowledged as a developmental necessity. India's low per capita energy use for space cooling (69

States	SDG7	SDG6	SDG12	SDG16	SDG11	SDG3	SDG10	SDG1	SDG15	SDG8	SDG4	SDG2	SDG13	SDG5	SDG9	Composite SDG
Kerala	100	89	65	80	75	72	69	83	77	62	80	80	69	61	60	75
Himachal Pradesh	100	85	77	73	79	78	78	80	68	78	74	62	62	62	61	74
Tamil Nadu	100	87	78	71	79	81	74	86	60	71	69	66	61	59	71	74
Andhra Pradesh	100	92	84	77	78	77	74	81	69	67	60	62	63	58	62	72
Goa	100	100	47	61	89	72	75	83	59	76	71	78	44	65	68	72
Karnataka	100	85	89	76	78	78	67	68	67	66	64	52	62	57	64	72
Uttarakhand	100	85	82	86	76	77	77	74	64	63	70	61	60	46	56	72
Sikkim	100	89	76	72	85	62	61	80	73	71	58	69	65	58	52	71
Maharashtra	100	90	82	69	87	83	71	66	62	62	64	44	58	61	66	70
Gujarat	94	93	60	82	87	86	64	66	61	64	52	46	67	49	72	69
Telangana	100	96	73	71	76	67	67	68	81	73	63	60	43	41	59	69
Mizoram	100	85	87	81	51	79	64	80	48	51	60	72	66	54	32	68
Punjab	100	66	71	76	91	77	68	69	48	57	60	73	51	45	69	68
Haryana	100	80	77	71	81	72	68	69	48	59	64	58	51	43	66	67
Tripura	83	82	99	80	67	67	85	82	69	57	42	52	41	39	35	65
Manipur	96	87	89	69	65	68	70	60	60	36	63	64	57	41	35	64
Madhya Pradesh	86	88	78	66	81	62	51	44	84	60	45	43	49	55	37	62
West Bengal	98	81	79	81	45	76	71	59	53	57	54	46	39	41	53	62
Chhattisgarh	78	89	64	71	78	60	72	49	65	64	58	37	38	64	36	61
Nagaland	69	87	91	79	48	61	46	73	63	48	39	64	69	48	30	61
Odisha	80	86	73	58	70	67	66	41	83	48	45	42	70	46	46	61
Arunachal Pradesh	85	67	77	64	39	64	69	54	93	56	41	66	58	37	31	60
Meghalaya	56	75	73	72	61	70	88	77	64	63	48	37	62	61	25	60
Rajasthan	100	64	74	73	81	70	45	63	43	57	60	53	45	39	45	60
Uttar Pradesh	100	83	79	79	77	60	41	44	61	53	53	41	39	50	42	60
Assam	98	64	66	62	56	59	68	62	78	50	43	41	53	25	39	57
Jharkhand	77	83	56	70	71	74	65	36	71	54	45	19	25	51	37	56
Bihar	78	91	50	73	67	68	48	32	62	50	29	31	16	48	24	52
UNION TERRITORIES	SDG7	SDG6	SDG12	SDG16	SDG11	SDG3	SDG10	SDG1	SDG15	SDG8	SDG4	SDG2	SDG13	SDG5	SDG9	Composite SDG
Chandigarh	100	99	78	73	98	74	100	75	85	70	79	97	61	58	45	79
Delhi	100	61	56	62	68	90	72	81	81	65	75	63	55	33	66	68
Lakshadweep	83	100	63	77	56	78	75	61	67	62	62	74	68	58	40	68
Puducherry	98	91	66	86	76	70	62	75	56	68	70	59	23	66	59	68
Andaman and Nicobar Islands	100	87	73	46	85	68	67	71	72	53	57	45	77	68	23	67
Jammu and Kashmir	100	88	95	74	57	70	65	69	62	47	49	71	63	46	42	66
Ladakh	100	84	95	74	57	70	65	79	27	50	49	71	66	46	48	66
Dadra and Nagar Haveli & Daman and Diu	71	95	62	75	89	80	66	65	62	57	56	27	18	63	47	62
India	92	83	74	74	79	74	67	60	66	61	57	47	54	48	55	66
Target	100	100	100	100	100	100	100	100	100	100	100	100	100	100	100	100

Aspirant (0-49) Performer (50-64) Front Runner (65-99) Achiever (100)

Fig. 3 SDG goal-wise performance of different Indian states. *Source* NITI Aayog SDG India Index and Dashboard 2020–21 (NITI Aayog, 2024)

kWh) compared to the global average of 272 kWh indicates that it has one of the lowest accesses to cooling in the world (Fig. 5).

In India, only around 8% of homes use air conditioners, which is increasing at an average rate of 12%. Domestic refrigerators are roughly 5% more common.

High levels of urbanization and limited access to cooling are characteristics of India's rapidly expanding economy. In order to satisfy people's essential needs and justifiable desires, such as a comfortable living and working environment for

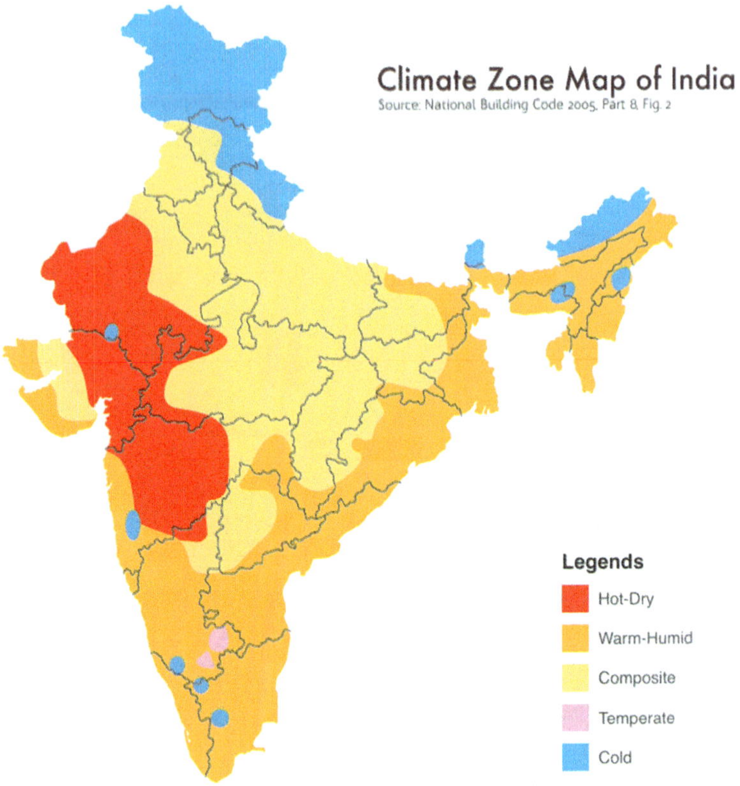

Fig. 4 Climate classification of India

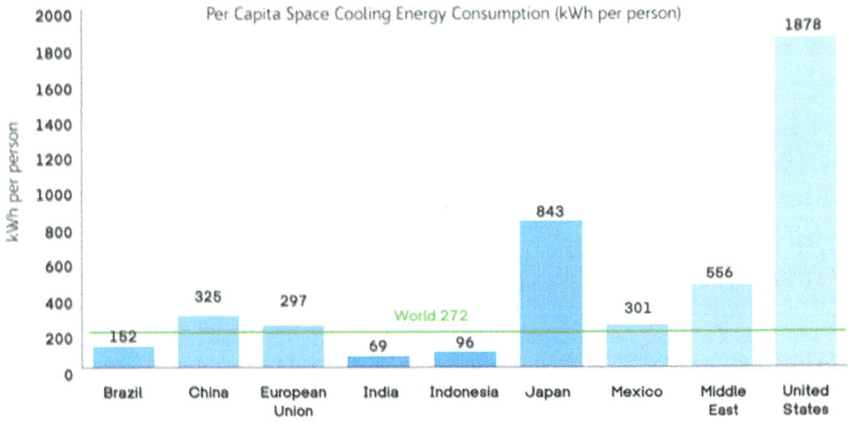

Fig. 5 Per capita space cooling energy consumption. *Source* Ozone Cell (2023)

increased productivity, health, and well-being, cooling demand is predicted to rise in line with the expanding economic and infrastructure development. By 2037–2038, the total cooling demand in the country is expected to increase eight times over the baseline of 2017–2018. In comparison with the baseline, the cooling demand in the building sector has increased the highest, approximately eleven times; the cold chain and refrigeration sectors will expand about four times, and the transport air-conditioning sector will grow around five times by 2037–2038 compared to 2017–2018 levels.

9.3 *Space Cooling in Buildings*

The construction industry is a vital component of the national economy and is intrinsically linked to the country's overall development. As a significant consumer of energy, particularly for cooling systems in urban areas, the sector plays a pivotal role in shaping the built environment. With rapid urbanization, the industry is poised for substantial growth, with an estimated 70% of buildings yet to be constructed (IFC, 2017). Subsequently, the demand for space cooling, particularly for air-conditioning and refrigeration, is projected to increase. Nevertheless, the building sector offers substantial potential for mitigating cooling demand and enhancing energy efficiency, thereby contributing to reduced energy consumption, refrigerant use, and associated direct and indirect greenhouse gas (GHG) emissions.

Residential Building Sector: In 2017, India had approximately 272 million households. This number is projected to increase to 328 million by 2027–28 and 386 million by 2037–38. Additionally, the penetration of refrigerant-based cooling in residential buildings was 8% in 2017, and it's expected to reach 21% and 40% by 2027–28 and 2037–38, respectively.

Commercial Building Sector: In 2017–18, India's commercial building sector (including hospitals, hotels, retail, offices, educational institutions, assembly places, transit buildings, and warehouses) had a floor area of 1160 million square meters. This is expected to grow 1.6 times by 2027–28 to 1880 million square meters and 2.7 times by 2037–38 to 3090 million square meters. Commercial buildings have a significantly higher air-conditioning demand than residential buildings (Fig. 6).

India has been seeing more frequent and severe heat waves along with a gradual increase in temperature that has pushed the start of summer forward by at least one month. Over the past ten years, peak electricity demand has increased by an average of 4% per year due to rising daily average temperatures and an increase in air conditioner ownership. There is a noticeable increase in cooling demand as average daily temperatures rise above 25 °C (Fig. 7), which is reflected in the GHG emissions (Fig. 8).

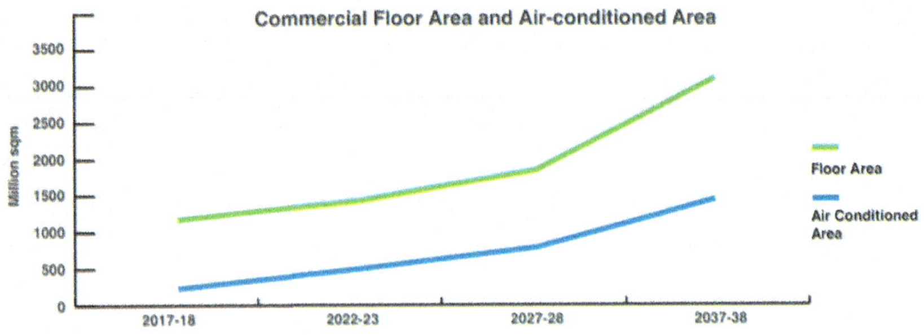

Fig. 6 Projected demand for air-conditioned space in India. *Source* Ozone Cell (2023)

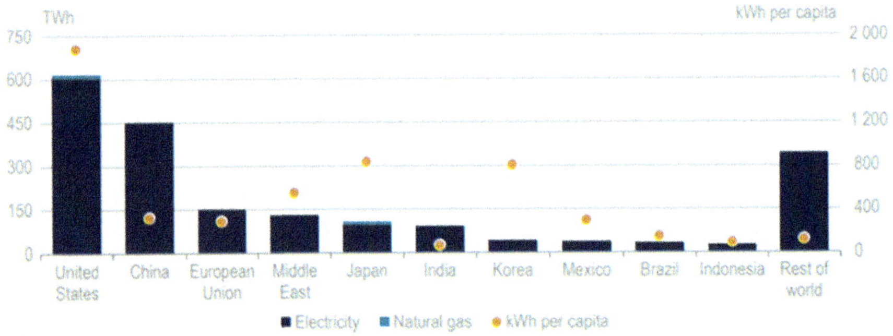

Fig. 7 Energy use for space cooling, 2016. *Source* Ozone Cell (2023)

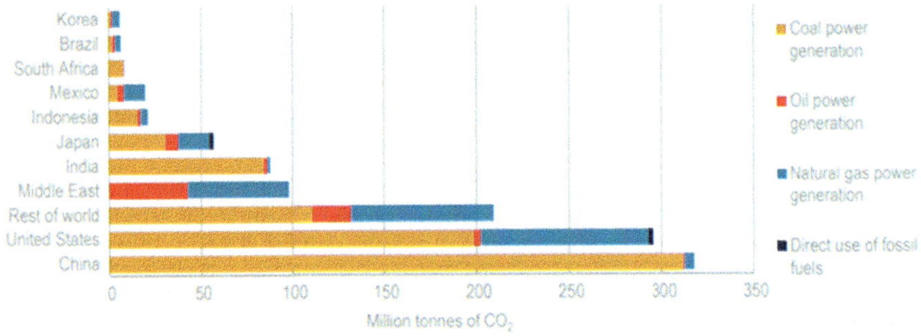

Fig. 8 CO_2 Emissions due to energy consumption for space cooling, 2016. *Source* Ozone Cell (2023)

10 Importance of Passive Cooling in India

The rapidly increasing cooling demand results in high energy consumption due to increased air conditioning. This creates a necessity for more power generation which leads to higher carbon emissions. Passive cooling can reduce impact on global warming. Urban heat islands, the phenomenon where dense urban areas experience elevated temperature increase health risk and energy consumption.

Passive cooling strategies such as optimum orientation, shading devices, provision for natural ventilation, green roofs and walls, courtyards, and use of local materials can increase comfort and thereby reduce cooling demands.

The building envelope, including roofs, floors, walls, doors, windows, and foundations, significantly influences cooling and ventilation needs. Material choices are crucial. Many countries are shifting from traditional heavy materials like stone and brick to lighter options like wood, aluminum, and composites, which may not provide the same level of insulation.

11 Studies on Passive Strategies in This Book

Passive cooling strategies is the only way to enhance indoor thermal comfort in warm-humid climate. This book is my third attempt in bring together research work done using field study data in the context of sustainable architecture. There have been several books on Sustainable Architecture, few on Sustainable Architecture in India, and hardly any on sustainable architecture which is specific to a particular context, climate, microclimate and situation. This book is one such attempt to understand the empirical outcome of the theories that are already known. Each of the chapters is a result of systematic research conducted on-site after the identification of a research gap through the latest review of the literature and present new knowledge which can help in making performance-based design decisions. These decisions align with making India Greener and sustainable.

This book deals with primary field-based assessment of the impact of passive strategies to enhance indoor thermal performance in India. The second chapter examined the potential and challenges of implementing energy retrofit policies in developing nations, specifically focusing on India's existing building stock. Through a hybrid methodology that combined a systematic literature review with scenario-based analysis, the research identified significant barriers to retrofit adoption, including regulatory, economic, and sociocultural constraints.

Chapter 3 deals with the indoor thermal performance of roof using Kadapa stone, a vernacular material in hot semi-arid region in India. The study found that RCC roof slabs is not an appropriate choice for roof construction in hot semi-arid conditions as it did not perform well thermally as compared to the Kadapa-lime roof. Vernacular buildings with local materials like Kadapa stone, lime concrete roof, and other passive strategies helped in achieving close to comfort conditions during the summer time in the hot semi-arid climate. Chapter 4 analyses the environmental footprint of cross-laminated timber (CLCT) as a building material in India. The study finds that the embodied energy of the CLCT building is 37% lesser than the RCC building. Further, the embodied greenhouse gas of the CLCT building is 186% lesser than a similar RCC building. Overall, in terms of embodied energy and embodied greenhouse gas consumption, CLCT building outperformed the RCC building.

Chapter 5 analyses the life cycle cost and CO_2 footprint of two affordable green homes in India. The two case studies are analyzed for their carbon footprint and life cycle analysis and compared their sustainable performance with the conventional buildings. It is found that Building 1, an independent home constructed with CSEB, in a temperate climate has over 78% reduction in the carbon footprint with over 50% life cycle cost compared to conventional building.

Chapter 6 demonstrates the importance of integrating climate-responsive design strategies at both the micro (dwelling unit) and macro (layout and street level) scales to enhance indoor comfort and reduce energy consumption in affordable housing in India. The results indicate that the layout with the main street oriented along the NE–SW direction performs approximately 10% better in winter, while in summer, the N–S street orientation outperforms the base case with an E–W orientation. Among the different building morphologies, the shorter typology proves to be the most beneficial, offering a 0.15% cooler environment compared to the base case, when compared to U-shape, square, and longer façade typologies. In terms of sky view factor (SVF) and aspect ratio, the base case, with a building height of 12 m and street width of 12 m, performs better in most instances than the other alternatives. Lastly, the case with the most vegetation shows superior performance over all other vegetation alternatives.

Chapter 7 demonstrates the impact of integrating of biophilic design principles in development control regulations with a case of educational and office building in India. The study showed that integration of small waterbody, good natural views, calm zones, aromatic spaces, functional living walls and greenery, biophilic form and façade, colorful interiors, easy accessibility, green outdoor spaces for leisure activities enhance the user perception significantly in an educational campus. The study also reveals that it is imperative to integrate biophilia strategies into development control regulations to create sustainable and resilient communities paving the way toward achieving the goals of SDG 11.

A comparative study of the green residence in Chennai and the vernacular building in Kanchipuram is assessed in Chap. 8. The green residence demonstrates superior energy efficiency with over 50% lower kWh/sqm consumption. On the other hand, the vernacular building, with its higher thermal mass and traditional materials, experiences a higher temperature difference of 10.4 °C than 6.2 °C in the green building. Chapter 9 discusses the significance of orientation to wind-flow in residential complexes for Chennai. The results indicate that the more suitable block orientation is between + 30° and + 45° tilted to southeast to capture prevailing south-west annual winds to increased velocity to ventilate the interior spaces. The long axis should be oriented between + 45° and + 60° from north axis. A comparative analysis of a baseline and remodeled scenario is studied to understand the presence of Green

Infrastructure to Mitigate Heat Intensity in Chap. 10. These results indicate that for microlevel Green Infrastructure interventions, physiographic conditions may also be considered. Instead of providing centralized solutions, decentralized solutions might be more effective. The research found that green infrastructure interventions were successful in transforming cities into more environmentally friendly and creative models. This approach has considerable promise in mitigating the UHI impact in urban settings.

Studied such as these is an urgent necessity in the face of a changing climate in a developing country like India, where the number of unbuilt buildings is burgeoning. The path to a sustainable future for India's built environment lies in integrating traditional design wisdom with modern innovations, promoting energy efficiency, and reducing the carbon footprint of buildings. As the world moves toward Net-Zero emissions, India's role will be crucial, not only as a rapidly developing nation but also as a key player in global sustainability.

References

Ahluwalia, M. S., & Patel, U. (2021). *Getting to net zero: an approach for India at CoP-26*. Centre for Social and Economic Progress.

Carboncopy. (2019). *India's cooling action plan: Losing steam amidst surge in temperature*. AC sales. https://carboncopy.info/indias-cooling-action-plan-losing-steam-amidst-surge-in-temperature-ac-sales

Dhanapal, G., Gupta, D., & Prakash, A. (2023). Barriers and opportunities in achieving climate and sustainable development goals in India: A multilevel analysis. *Journal of Integrative Environmental Sciences, 20*(1), 1–16. https://doi.org/10.1080/1943815X.2022.2163665

Dixit, H., Bhasin, S., Bhattrai, P., Mani, S., Laha, A. K. (2023). *Perceptions of refrigeration and air-conditioning service technicians on India cooling action plan: Enhancing servicing sector jobs for a just transition*. New Delhi: Council on Energy, Environment and Water

IFC. (2017). Climate Investment Opportunities in South Asia report, International Finance Corporation, World Bank Group, Creating Markets, Creating opportunities as accessed from https://www.ifc.org/content/dam/ifc/doc/mgrt/17663-ifcindia-factsheet-v3.pdf on 23rd Feb 2025.

Indian Green Building Council. (2020). *IGBC Green building rating systems*. Indian Green Building Council. Retrieved from https://igbc.in

Intergovernmental Panel on Climate Change (IPCC). (n.d.). *Climate change: The IPCC 1990 and 1992 assessments*. IPCC. Retrieved November 20, 2024, from https://www.ipcc.ch/report/climate-change-the-ipcc-1990-and-1992-assessments/

Iqbal, S., Sarfraz, M., Ayyub, M., Tariq, M., Chakrabortty, R., Ryan, M., & Alamri, B. (2021). A comprehensive review on residential demand side management strategies in smart grid environment. *Sustainability, 13*(17), 7170. https://doi.org/10.3390/su13137170

Gamero-Salinas, J., Monge-Barrio, A., Kishnani, N., López-Fidalgo, J., & Sánchez-Ostiz, A. (2021). Passive cooling design strategies as adaptation measures for lowering the indoor overheating risk in tropical climates. *Energy and Buildings, 252*, 111417. ISSN 0378-7788. https://doi.org/10.1016/j.enbuild.2021.111417

Khosla, R., Miranda, N. D., Trotter, P. A., et al. (2021). Cooling for sustainable development. *Nature Sustainability, 4*, 201–208. https://doi.org/10.1038/s41893-020-00627-w

Lessio, M., Byers, E., Pachauri, S., & Rao, N. D. (2019). Improving the SDG energy poverty targets: Residential cooling needs in the Global South. *Energy and Buildings, 186*, 405–415. ISSN 0378-7788.

Ministry of Environment, Forest & Climate Change. (2022). *Passive & low energy cooling strategies for achieving thermal comfort in India's upcoming affordable housing*. https://aeee.in/our-publications/passive-low-energy-cooling-strategies-for-achieving-thermal-comfort-in-indias-upcoming-affordable-housing/

Miranda, N. D., et al. (2023). Change in cooling degree days with global mean temperature increasing from 1.5 °C to 2.0 °C. *Nature Sustainability*, 13 July 2023. https://doi.org/10.1038/s41893-023-01155-z

MoEFCC. (2024). *Passive cooling strategies for sustainable buildings*. https://ozonecell.nic.in/wp-content/uploads/2024/09/5-Passive-Cooling-Strategies-for-Sustainable-Buildings.pdf

NITI Aayog. (2024). *SDG India index 2023–2024*. https://www.niti.gov.in/sites/default/files/2024-07/SDG_India_Index_2023-24.pdf

Ozone Cell. (2023). India's journey towards sustainable cooling, LiFE, lifestyle for environment, Ministry of Environment, Forest and Climate Change, Government of India.

Panchabikesan, K., Vellaisamy, K., & Ramalingam, V. (2017). Passive cooling potential in buildings under various climatic conditions in India. *Renewable and Sustainable Energy Reviews, 78*, 1236–1252.

United Nations Department of Economic and Social Affairs (UN DESA). (2024, June). *The sustainable development goals report 2024*. https://sdgs.un.org/goals

United Nations Framework Convention on Climate Change (UNFCCC). (n.d.). *The Kyoto protocol*. UNFCCC. Retrieved November 20, 2024, from https://unfccc.int/documents/2409

Energy Retrofit Policies: Global Challenges and Insights for India's Path to Net-Zero

Shalini Keshri and Priyanka Dey

Abstract The global energy sector faces significant challenges despite notable cases and advancements. The notion of energy retrofit is still unfamiliar to many nations, resulting in limited development. One of the major challenges has been identified in the planning and implementation of energy policies. In India, the concept of energy retrofitting has not been considered as a potential solution for addressing the growing energy requirements of the current building inventory. Residential buildings are primary users of energy. Exploring energy sustainability for these structures is crucial in order to obtain a net-zero status. This study seeks to assess cases on energy policy planning for retrofit of existing buildings, undertaken by 21 nations. Drawing from existing literature, four main barriers to energy policy planning and implementation were identified. Further a hypothetical scenario-based assessment for energy demands between 2021 and 2071 was performed, based on projections for population and housing stock. The findings of the analysis emphasized the importance of quantifying retrofit potentials to support India's net-zero ambitions. Findings indicate that a phased, tailored policy framework could help India achieve sustainable energy goals while effectively addressing identified challenges and barriers.

Keywords Energy policy · Energy retrofit · Net-zero · Building sustainability · Policy barriers · Scenario-based assessments

1 Introduction

The adverse effects of climate change became more noticeable around the turn of the millennium. This prompted a surge in the formulation of global and national energy policies. However, in most cases, the desired pace of progress was not achieved.

S. Keshri (✉) · P. Dey
Department of Architecture and Regional Planning, Indian Institute of Technology Kharagpur, Kharagpur, West Bengal 721302, India
e-mail: shalini.keshri.96047@kgpian.iitkgp.ac.in

P. Dey
e-mail: priyanka@arp.iitkgp.ac.in

J. Vijayalaxmi (ed.), *Sustainable Building Performance*,
https://doi.org/10.1007/978-981-96-2428-7_2

The Paris Agreement of 2016 marked the significant need to provide immediate control on energy sector and future energy concerns at an international consensus (UNFCCC, 2016). Since then, improving energy efficiency has become a global priority, leading to a shift in energy sector practices—ranging from the choice of materials and construction methods to the integration of new technologies. At COP26 in 2021, India pledged to achieve net-zero emissions by 2070 (The United Nations Climate Change Conference, 2021).

Despite these efforts, energy demand has continued to rise. The Energy Development Index, defined by the International Energy Agency (IEA), indicates that nearly 10% of the global population still lacks access to reliable energy and is classified as "*energy poor*" (International Energy Agency, 2022b). Additionally, energy consumption has grown significantly across sectors, particularly in the building sector, where final energy consumption increased by 26.77%, between 2446 Mtoe[1] in 2000 and 3101 Mtoe in 2018 alone. Projections suggest a further rise in demand by 11.41% by 2030 and 21.18% by 2040 under current policy scenarios. Moreover, global electricity consumption is expected to increase from 19% in 2019 to 23% by 2040 (International Energy Agency, 2019). The building sector alone accounts for approximately 40% of global energy consumption and over 50% of global electricity demand (Ala-Juusela & ur Rehman, 2020; International Energy Agency, 2022a; Pyloudi et al., 2015; T.E.R.I, 2013). These trends underscore the escalating energy demand and highlight the pressing need for effective measures to address it (Dey & Keshri, 2023). However, existing policies have struggled to bridge the widening gap between energy demand and efficiency, creating a paradoxical situation.

Within the building sector, commercial and residential stocks are the primary energy consumers. A keyword search of the Scopus and Web of Science databases[2] indicates that discussions around "studies on energy retrofit" are more frequent for commercial buildings compared to residential structures. In India, residential energy consumption increased by 3% between 2006 and 2019, emphasizing the need for a greater focus on residential buildings to improve energy performance (NITI Aayog, Government of India, 2019). As new buildings continue to be constructed, it is crucial to acknowledge that nearly 75% of the existing building stock will still be in use in the 2030s and 2040s. Therefore, focusing on improving the energy performance of existing structures is essential to meet future energy needs. Although efforts to enhance new construction practices have made progress, the limited timeframe to meet energy goals makes it clear that focusing solely on new buildings is insufficient. Consequently, there is growing attention on energy retrofits for existing structures.

Addressing the rising energy demand through strategic planning and policy implementation for the existing building stock presents a critical opportunity to bridge the gap between energy consumption and efficiency. This is especially important for meeting climate change and sustainable development goals. As developing nations

---

[1] Mtoe-Million tons of oil equivalent.

[2] A keyword analysis conducted on the Scopus and Web of Science databases on February 7, 2024 identified 4554 research publications with the keyword 'Energy retrofit for existing buildings' of which only 1269 articles addressed energy retrofit for residential buildings.

like India continue to urbanize rapidly while pursuing major energy goals, it is crucial to carefully evaluate each step in the process (Dey & Keshri, 2023). Although numerous studies emphasize the potential of energy retrofits to address energy challenges, the availability of practical, evidence-based case studies remains limited. This lack of practical examples heightens hesitation and reluctance to engage in retrofitting projects. Furthermore, the absence of clear assessments and quantifications of potential benefits compounds the reluctance to invest in such projects and policies.

Currently, only a limited number of nations possess the economic and developmental stability needed to make meaningful contributions through energy retrofits as part of their sustainable development efforts. Very few countries have managed to implement energy retrofit policies capable of achieving even a 1% annual retrofit rate. However, projections made by the International Energy Agency in the context of energy retrofit measures suggest the need to achieve at least a 5% rate of retrofit per annum and to retrofit at least 20% of the existing structures by 2030, to satisfy global targets (International Energy Agency, 2021). The trend of energy retrofit in existing buildings is a relatively recent notion that is being contested in terms of its application as well as the level of efficiency; and user satisfaction that can be achieved from it (Dandia et al., 2021; Kumbaroglu & Madlener, 2012; Menassa, 2011; Surya Prakash et al., 2021).

This research aims to shed light on these issues and offer insights into energy retrofit policy planning for India. Section 2 outlines the methodology used in this study. Section 3 presents the results and discusses the potential challenges and contributions of energy retrofit policies in advancing India's progress toward net-zero targets. Section 4 concludes the study.

2 Methodology

The initial literature review and background on energy retrofit policy planning and implementation highlight that while energy retrofits are recognized as a viable solution, their actual adoption remains limited due to various hesitations. To better understand energy policies and assess their potential in developing nations like India, a hybrid approach combining both qualitative and quantitative research is essential. To fulfill this, the study's methodology was designed to address the following research questions (RQs):

RQ 1: What challenges and barriers would the developing countries like India face when planning and implementing energy retrofit policies for existing buildings?
RQ 2: If implemented, how can the benefits of these policies be quantified, and what should be the timeline for achieving them?

To explore these questions, the study methodology was divided into two phases (Fig. 1). Phase I consisted of a systematic literature review aimed at identifying the

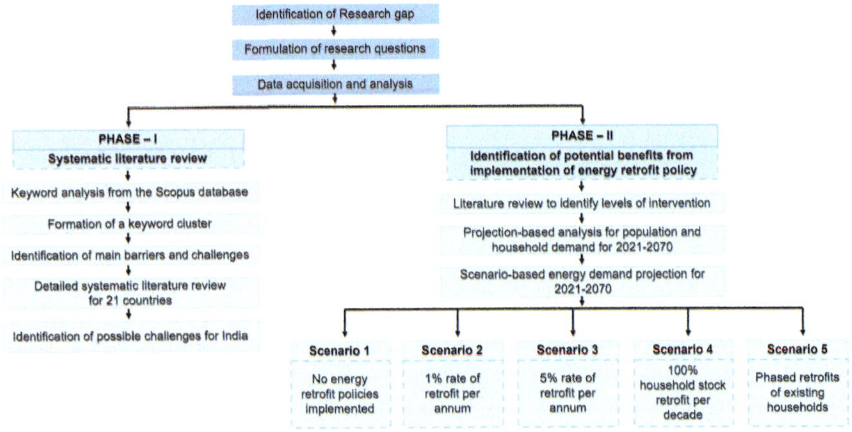

Fig. 1 Methodological flow adapted for this study. *Source* Author

key barriers and challenges faced in energy retrofit policy planning and implementation globally. Phase II involved a scenario-based analysis to quantify the potential energy savings from energy retrofits in India's residential building stock. A detailed discussion of both phases is provided in Sects. 2.1 and 2.2.

2.1 Phase I: Systematic Literature Review of Existing Energy Retrofit Policies Worldwide

The first phase of this study aimed to identify countries with existing energy retrofit policies and review the barriers and challenges they encountered during policy implementation. Using a keyword analysis in the Scopus database and VOS viewer (van Eck & Waltman, 2010), 42 keywords were mapped into nine clusters, and categorized into four main barriers (Fig. 2). This review covered 68 research publications from 21 countries, with a detailed summary presented in Tables 2 and 3, and the list of policies in Appendix.

The four main barriers identified from the literature review were:

- ***Regulatory or control barriers*** present significant challenges, shaping the approach to policy implementation, regularization, and overall effectiveness in addressing energy issues (Al Naqbi et al., 2019; Bertone et al., 2018a; Brown, 2014).
- ***Economic, market-based, or fiscal barriers*** primarily involve funding challenges, investment priorities, and the availability of incentives, taxation mechanisms, and market acceptability for retrofitting initiatives (Ahmed & Asif, 2021; Hrovatin & Zorić, 2018; Huang et al., 2016; Liu et al., 2020).

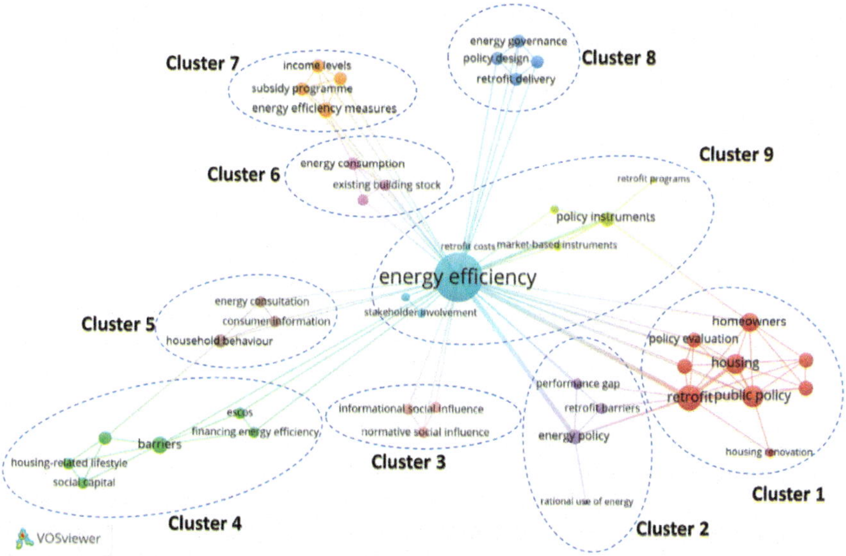

Fig. 2 Keyword clusters for challenges and barriers related to energy policies across the world. *Source* Author

- ***Socio-economic barriers*** encompass factors such as awareness, affordability, demographic influences, willingness to invest, and the complexities of ownership in energy retrofit decisions (Hrovatin & Zorić, 2018; Jovović et al., 2023; Pombo et al., 2019).
- ***Information barriers*** refer to the gaps in knowledge, data reliability, capacity building, and public involvement that hinder effective implementation of energy retrofit policies (Brown, 2014; Dolšak et al., 2020; Huang et al., 2016).

2.2 *Phase II: Identification and Quantification of Potential Benefits from Energy Retrofit Policies*

Studies specific to India have noted that our building energy demand and consumption patterns are distinct from other countries, rendering standard international solutions ineffective in the Indian context (Chaturvedi et al., 2014). Given that most existing residential buildings in India do not undergo retrofits or renovations during their lifetime, significant potential for energy savings remains untapped. Therefore, considering this gap in the research, a scenario-based analysis approach was taken to quantify the potential energy savings from energy retrofits in India's residential building stock.

Five scenarios (S) were hypothesized (Table 1), to evaluate the potential impact of energy retrofits on India's housing stock. Population projections from 2021 to 2071 were made, based on India's 1991–2011 census data, assuming an average household size of 4.8 (MOSPI, 2021). The analysis categorized housing stock into three age-based conditions: "Good" (less than 40 years), "Liveable" (40–80 years), and "Dilapidated" (over 80 years) (Census of India, 2011), with different levels of retrofit interventions—minor, moderate, and major—resulting in minimum expected energy demand reductions of 16.70%, 19.87%, and 24.12%, respectively (Sharma et al., 2022). To calculate the energy demand under different scenarios, household stock projections and base year energy demand for 2021 were estimated. Projections were made by analyzing residential energy consumption data from 2009 to 2019, identifying annual increments, and linearly extrapolating to predict demand for 2021–2071.

$$\mathrm{ED}_{t+n} = \mathrm{ED}_t + (n * I) \tag{1}$$

where ED_{t+n} is the energy demand for a projected year, ED_t is the energy demand for the base year, n is the number of years from the base year (for a decade, $n = 10$), and I is the average annual increment identified from historical data. The population was projected using arithmetic, geometric, and incremental methodologies, with the average used as the final estimate.

$$P_{t+n} = \frac{\sum\left(P_{\mathrm{Arithmetic},t+n} + P_{\mathrm{Geometric},t+n} + P_{\mathrm{Incremental},t+n}\right)}{3} \tag{2}$$

where P_{t+n} is the projected population for a given decade, and $P_{\mathrm{Arithmetic},\, t+n}$; $P_{\mathrm{Geometric},\, t+n}$; $P_{\mathrm{Incremental},\, t+n}$ are the results of different projection methodologies. The average energy demand per household for each decade was then calculated. Here, $\mathrm{ED}_{\text{per household, per decade}}$ is the decadal energy demand per household, ED_t is the existing or projected energy demand for base (2011)/future year (2021–2071), and

Table 1 Scenarios hypothesized to quantify the potentials of energy retrofit policy implementation in India

Scenario code	Scenario assumption
S1	No energy retrofit efforts for existing residential structures. Future energy demand for 2021–2071 was projected using Eq. (1)
S2	1% of the total housing stock is retrofitted annually, meaning 10.46% of the total stock is retrofitted each decade. Various intervention levels (Minor, Moderate, Major) were explored using Eq. (4)
S3	5% of the housing stock is retrofitted annually, with 62.89% retrofitted per decade. Intervention levels similar to *S2* were applied (Eq. 4)
S4	The total housing stock in each decade (2021–2071) is subjected to different intervention levels (Eq. 4)
S5	Phase-wise interventions were applied to housing stock based on age and condition (Good, Liveable, Dilapidated) (Eq. 7)

HH_t is the existing or projected household stock for the base year (2011)/future year (2021–2071).

$$ED_{\text{per household,per decade}} = ED_t/HH_t \tag{3}$$

Projections for energy demand under Scenarios 2, 3, and 4 were made by calculating the energy savings from retrofitted and non-retrofitted households, based on the rate of retrofitting and the percentage of energy reduction achieved through various interventions.

$$ED_{S_y} = ED_{\text{Retrofitted},x} + ED_{\text{Not-retrofitted}} \tag{4}$$

ED_{S_y} is the energy demand under different scenarios, 'S_y', where $y =$ Scenario 2, 3 or 4. $ED_{\text{Retrofitted},\,x}$ and $ED_{\text{Not-retrofitted}}$, considering different levels of interventions are:

$$ED_{\text{Retrofitted},x} = \left[\{(P_r/100) * H_t\} * ED_{\text{per household, per decade}} * (1 - x)\right] \tag{5}$$

$$ED_{\text{Not-retrofitted}} = HH_{\text{Not-retrofitted}} * ED_{\text{per household, per decade}} \tag{6}$$

The P_r represents the cumulative or decadal percentage of the housing stock being retrofitted, which can be 10.46%, 62.89%, or 100%, depending on the scenario. H_t refers to the total housing stock in a given year (e.g., 2021, 2031, 2041, up to 2071). The variable 'x' denotes the percentage of energy demand reduction achieved through various levels of retrofit interventions, with reductions of 16.70% for minor interventions, 19.87% for moderate interventions, and 24.12% for major interventions.

For Scenario 5, a phase-wise approach was adopted, with the housing stock subjected to retrofitting based on its age, and assumptions about the type of interventions applied as the stock aged:

$$ED_t = \sum_{A_t}\left[\left\{P_{\text{Retro}\,(A_t, R_{\text{type}})} * E_{A_t, R_{\text{type}}}\right\} + \left(D_{A_t} * E_{A_t,\text{Dilapidated}}\right)\right] \tag{7}$$

Here, ED_t is the energy demand for a particular year "t," A_t is the age of the building stock, calculated from the base year 1991. D_{A_t} is the percentage of household stock considered to be dilapidated at age A_t. The energy demand for a household stock at age A_t is represented as $E_{A_t,R_{\text{type}}}$, where R_{type} refers to the type of retrofit measure applied at time "t," such as minor, moderate, or major retrofits. Similarly, $E_{A_t,\text{Dilapidated}}$ refers to the energy demand for households reaching energy obsolescence at age A_t. The percentage of household stock subjected to a particular retrofit measure is denoted as $P_{\text{Retro}(A_t,R_{\text{type}})}$ with different levels of intervention applied based on the building's age.

For newer buildings (0–30 years), no retrofit intervention was assumed. As buildings reach the age of 30–40 years, it was assumed that around 62.89% of the existing households (5% rate of retrofit per annum) undergo minor retrofits. Similarly, when buildings are 40–50 years old, moderate retrofits interventions were applied to the majority (62.89%), with minor retrofits for the remainder. For buildings aged 50–60 years, it was assumed that a small percentage (10.46% cumulatively, per decade, considering a 1% rate of retrofit per annum) undergoes major retrofits, while the rest existing stock received moderate or minor interventions. From 60 to 70 years onward, retrofits become more extensive, with 62.89% receiving major retrofits. Once buildings reach 70–80 years of age, energy obsolescence begins, with 5% of the stock classified as dilapidated, but the remaining 95% still undergoes major retrofits. The proportion of dilapidated buildings increases to 10% for buildings aged 80–90 years, and 20% for those aged 90–100 years, with major retrofits applied to the remaining stock in each case. This approach allowed a phased, age-dependent retrofit strategy that maximizes coverage of household stock while addressing the degradation of building stock over time. The outcomes of the various scenarios indicate differing levels of potential, are analyzed and discussed in the following section.

3 Results and Discussion

Existing literature suggests that while energy retrofit projects provide a practical solution to energy concerns, they face low acceptance in common practice. Instead, a "dispose and discard" approach is often seen as more convenient. This hesitation to pursue energy retrofits in existing buildings stems from various challenges and barriers documented in numerous studies over time. This section reviews insights from the current energy retrofit policies implemented worldwide, examining how different countries address these challenges.

3.1 Review of Policies Across the World

The types of policies implemented across the different regions vary widely, reflecting on their different political, socio-cultural, and economic settings. As a part of the assessment for RQ1, the four main barriers identified through the policy implementation experience from 21 countries (Appendix) were compared to India, where each region displayed the following challenges (Table 2). The 21 countries identified for the case study were grouped into five geographical regions: The European Union (EU), Oceania, The Gulf Cooperation Council (the GCC), Asia, and the Americas.

Table 2 Comparison of Regulatory and Economic-Market-Fiscal barriers identified by case studies across various countries

Barrier	Challenge	Countries considered for case study								
		The European Union							Oceania	
		U.K.	Germany	Spain	Slovenia	Bosnia and Herzegovina	Scotland	Croatia	Australia	New Zealand
		1	2	3	4	5	6	7	8	9
Regulatory or control barriers	Policy formulation		✔	✔				✔	✔	✔
	Policy governance approach: Top-down versus bottom-up	Top-down	Top-down	Top-down	Top-down	Top-down	Top-down	Top-down	Bottom-up	Top-down
	Policy implementation	National	National	National	National	National	National	National	Regional	National
	Partially successful cases		✔			✔		✔	✔	
Economic, market-based or fiscal barriers	Income category*	*H*	*H*	*H*	*H*	*U*	*H*	*H*	*H*	*H*
	Lack of funds			✔	✔	✔			✔	
	Priority of investments focused on old or new buildings	Both	Both	Both	Old	Old	Both	Both	Both	Both
	Incentives or Subsidies		✔			✔				
	Taxation	✔			✔		✔			

(continued)

Table 2 (continued)

Barrier	Challenge	Countries considered for case study								
		The European Union							Oceania	
		U.K.	Germany	Spain	Slovenia	Bosnia and Herzegovina	Scotland	Croatia	Australia	New Zealand
		1	2	3	4	5	6	7	8	9
	Market acceptability	✔	✔		✔	✔	✔			

Barrier	Challenge	Countries considered for case study												
		The GCC						Asia					America	
		Bahrain	Saudi Arabia	Kuwait	Oman	Qatar	U.A.E.	Japan	Singapore	China	India	South Korea	North America	Canada
		10	11	12	13	14	15	16	17	18	–	19	20	21
Regulatory or control barriers	Policy formulation	✔	✔	✔	✔	✔	✔	✔	✔	✔			✔	✔
	Policy governance approach: Top-down versus bottom-up	Top-down	Top-down	Top-down	Top-down	Top-down	Top-down	Top-down	Top-down	Bottom-up	Top-down	Top-down	Top-down	Top-down

(continued)

Table 2 (continued)

Barrier	Challenge	Countries considered for case study												
		The GCC						Asia					America	
		Bahrain	Saudi Arabia	Kuwait	Oman	Qatar	U.A.E.	Japan	Singapore	China	India	South Korea	North America	Canada
		10	11	12	13	14	15	16	17	18	–	19	20	21
	Policy implementation	National	National	National	National	National	National	Regional	Local	National	National	Regional	National	National
	Partially successful cases	✔	✔	✔	✔	✔	✔	✔	✔	✔			✔	✔
Economic, market-based or fiscal barriers	Income category*	*H*	*H*	*H*	*H*	*H*	*H*	*H*	*H*	*U*	*L*	*H*	*H*	*H*
	Lack of funds										✔			
	Priority of investments focused on old or new buildings	New	New	New	New	New	New	New	Both	New	New	Both	Both	Both
	Incentives or Subsidies							✔	✔	✔		✔		

(continued)

Table 2 (continued)

Barrier	Challenge	Countries considered for case study												
		The GCC						Asia					America	
		Bahrain	Saudi Arabia	Kuwait	Oman	Qatar	U.A.E.	Japan	Singapore	China	India	South Korea	North America	Canada
		10	11	12	13	14	15	16	17	18	–	19	20	21
	Taxation							✔	✔			✔		
	Market acceptability							✔	✔		✔	✔		

*Income category of countries: *H*, high income; *U*, upper middle income; *L*, lower middle income (UNCTAD, n.d.)

Table 3 Comparison of Socio-Economic and Information barriers identified by case studies across various countries

Barrier	Challenge	Countries considered for case study								
		The European Union							Oceania	
		U.K.	Germany	Spain	Slovenia	Bosnia and Herzegovina	Scotland	Croatia	Australia	New Zealand
		1	2	3	4	5	6	7	8	9
Socio-economic barriers	Lack of awareness			✔	✔	✔			✔	✔
	Effect of demographic variables: Age, income, educational level	✔	✔	✔	✔	✔	✔	✔	✔	✔
	Affordability for citizens	✔	✔	✔	✔	✔	✔	✔	✔	✔
	Lack of willingness	✔	✔	✔	✔		✔			
	Owner-tenant dilemma	✔	✔	✔	✔	✔	✔	✔	✔	✔
Information barriers	Experience and capacity building	✔		✔		✔	✔	✔	✔	✔
	Data monitoring and information validation	✔	✔	✔	✔	✔	✔		✔	✔
	Public participation in policy development		✔	✔	✔	✔		✔	✔	✔

(continued)

Table 3 (continued)

Barrier	Challenge	Countries considered for case study												
		The GCC						Asia					America	
		Bahrain	Saudi Arabia	Kuwait	Oman	Qatar	U.A.E	Japan	Singapore	China	India	South Korea	North America	Canada
		10	11	12	13	14	15	16	17	18	–	19	20	21
Socio-economic barriers	Lack of awareness	✓	✓	✓	✓	✓	✓		✓	✓	✓	✓		
	Effect of demographic variables: Age, income, educational level	✓	✓	✓	✓	✓	✓	✓	✓	✓	✓	✓	✓	✓
	Affordability for citizens	✓	✓	✓	✓	✓	✓	✓	✓	✓		✓		
	Lack of willingness	✓	✓	✓	✓	✓	✓					✓	✓	✓
	Owner-tenant dilemma	✓	✓	✓	✓	✓	✓	✓	✓	✓	✓	✓	✓	✓

(continued)

Table 3 (continued)

Barrier	Challenge	Countries considered for case study												
		The GCC						Asia					America	
		Bahrain	Saudi Arabia	Kuwait	Oman	Qatar	U.A.E	Japan	Singapore	China	India	South Korea	North America	Canada
		10	11	12	13	14	15	16	17	18	–	19	20	21
Information barriers	Experience and capacity building	✓	✓	✓	✓	✓	✓	✓		✓				
	Data monitoring and information validation	✓	✓	✓	✓	✓	✓	✓	✓	✓		✓	✓	✓
	Public participation in policy development	✓	✓	✓	✓	✓	✓	✓	✓	✓		✓		

3.1.1 Regulatory or Control Barriers

A comparative analysis of regulatory and control barriers across regions highlights significant differences in policy frameworks, governance structures, and implementation methods. Predominantly, top-down governance models are used, such as in India, where energy policies are directed at the national level. Regions like the European Union and the GCC benefit from standardized frameworks that guide member states in formulating cohesive energy policies. However, the literature stresses the need for policy adaptation to fit regional socio-economic contexts, as national-level directives may lack flexibility. In contrast, countries like Australia and Bosnia and Herzegovina adopt decentralized, bottom-up approaches that allow for responsive, context-specific policies. For India, where broad energy goals exist but targeted policies for existing buildings are limited, implementing localized solutions could greatly boost public awareness and engagement with energy retrofit programs, improving adoption rates and program success.

3.1.2 Economic, Market-Based, and Fiscal Barriers

Economic barriers to energy retrofit adoption often arise from limited evidence of cost savings, leading to hesitancy even in well-funded, high-income nations. In developing economies like India, the lack of financial incentives and limited public resources further dampens investment in retrofits. Where incentives exist, they are typically inadequate to drive widespread adoption. India's retrofit challenge also includes balancing investments between upgrading existing buildings and constructing new energy-efficient structures. In regions where retrofit policies and materials are still emerging, resources for new builds and retrofits may become stretched.

3.1.3 Socio-economic Barriers

Socio-economic barriers to energy retrofits include low awareness, cost concerns, and demographic disparities, hindering widespread adoption. Limited understanding of retrofit benefits reduces public interest and willingness to invest, with the "owner-tenant dilemma" complicating matters, particularly in rental sectors where funding responsibilities and benefits are unclear. Countries like Japan and Canada address these issues through frameworks and incentives that help share retrofit costs. However, in developing nations such as India, disparities in income and education mean that many residents cannot afford retrofits without specific subsidies and incentives.

3.1.4 Information Barriers

Effective data monitoring, validation, and public engagement are vital to the success of energy policies, directly influencing outcomes. Regions like the EU and Oceania have advanced data systems for tracking retrofit effectiveness, enabling accurate impact assessments. In contrast, India's data monitoring is still developing, limiting the comprehensive evaluation of retrofits. Public involvement is prominent in Europe and North America, where stakeholder input fosters trust and supports engagement. In India, improved data systems and greater public participation could enhance adoption rates for energy retrofit projects. Initiatives by the Bureau of Energy Efficiency (BEE) and Indian Green Building Council (IGBC) aim to build capacity and train professionals for retrofit projects, helping drive forward energy retrofit adoption.

This comparative analysis highlights that, despite challenges, India's journey toward implementing energy retrofits offers a chance to address an overlooked sector in its energy landscape. In light of India's 2070 net-zero targets, exploring alternatives and proactively tackling barriers yields valuable insights from current case studies. However, the lack of measurable data on potential energy savings could amplify socio-economic and informational challenges, subsequently discouraging investment. To bridge this gap, a scenario-based approach was used to estimate achievable outcomes through targeted retrofit policies.

3.2 Potential Energy Demand Reductions by the Implementation of Energy Retrofit Policies in India

This analysis evaluated five distinct scenarios under RQ2, each examining different levels of retrofit interventions while considering the age and condition of the existing housing stock to estimate potential energy savings through targeted retrofit policies. The findings suggest that a well-structured policy framework can deliver significant energy savings over time and provide a foundational guide for future policy development. A detailed discussion of each scenario and comparative results follows:

3.2.1 Scenario 1: No Retrofit

Scenario 1 assumed that the existing housing stock for the base years (1991–2011) and projected household stock for future years (2021–2071) are not retrofitted at all and their energy demand progress at the current rates.

The projected trends indicate an anticipated decadal increase in energy demand by approximately 55–58%, leading to a cumulative rise of around 956 Terawatts (TW) over the study period (Fig. 3). This consistent growth trajectory underscores the potential pressures on India's energy resources and infrastructure if energy efficiency improvements in the building sector are not implemented. Such pressures could

Fig. 3 Energy demand projected under Scenario 1 considering no retrofit interventions. *Source* Author

result in elevated greenhouse gas emissions and strain on natural resources, posing significant threats to India's development.

Given India's commitment to achieving net-zero emissions by 2070, the current scenario raises critical concerns about the sustainability of energy consumption patterns. As demand escalates, the risk of supply shortages and rising energy costs becomes increasingly likely, alongside negative environmental consequences. This scenario emphasizes the essential role of energy retrofitting in enhancing efficiency across existing households and infrastructure. It also provides a baseline for assessing various policy initiatives targeting energy retrofits for current and anticipated residential housing stock in India.

3.2.2 Scenario 2: 1% Rate of Retrofit Per Annum (10.46% of Households Retrofitted Per Decade)

The literature indicates that less than 1% of global household stock undergoes retrofitting annually (Berwal & Yadav, 2021; International Energy Agency, 2021). Given this, Scenario 2 models a 10.46% decadal retrofit rate, assuming mandatory compliance with the 1% annual target.

Compared to the previous scenario, this case provides modest reductions of 1.75% to 2.53% in projected energy demand for household stock. Across different retrofit levels—Minor, Moderate, and Major—the data underscores the significance of even small-scale efficiency measures in managing India's energy needs (Fig. 4). By 2071, energy demand could be reduced by 19–28 TW, demonstrating that even minimal retrofits, achievable with lower financial investment, can effectively ease the load on energy resources and support sustainable energy management.

3.2.3 Scenario 3: 5% Rate of Retrofit Per Annum (62.89% of Households Retrofitted Per Decade)

Similar to the previous case, this scenario assumes a 62.89% decadal retrofit rate, with 5% of household stock retrofitted annually. Analysis of different intervention levels shows that this approach could reduce energy demand by 10.5% to 15.1%, depending on retrofit intensity (Fig. 5). Studies on energy retrofits suggest that an

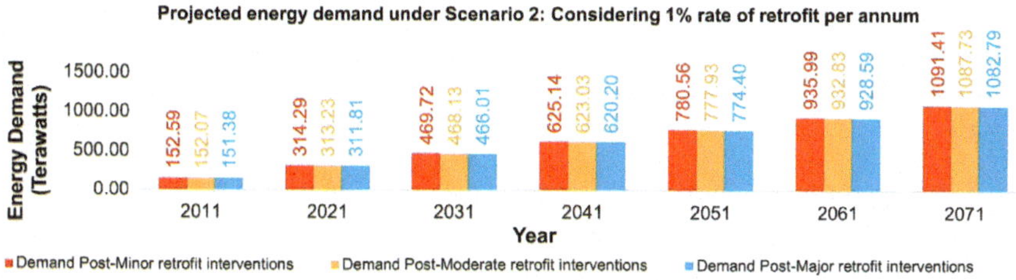

Fig. 4 Energy demand projected under Scenario 2 considering 1% rate of retrofit per annum. *Source* Author

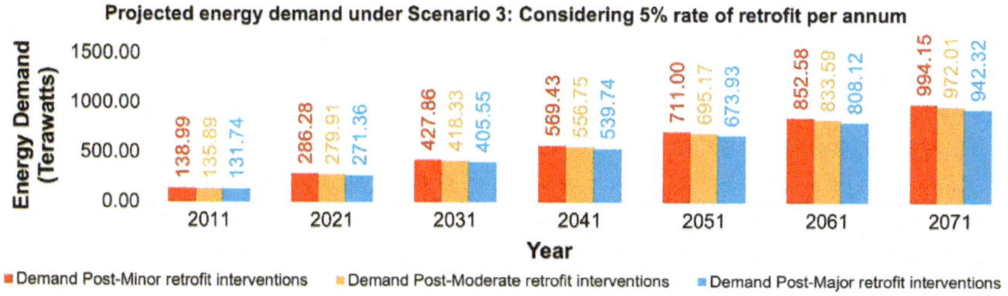

Fig. 5 Energy demand projected under Scenario 3 considering 5% rate of retrofit per annum. *Source* Author

annual retrofit rate of at least 5% is necessary to make a measurable impact on future energy demand. These findings highlight that high-intensity retrofitting is a viable policy approach to meaningfully reduce energy consumption, supporting a path toward sustainable energy management.

3.2.4 Scenario 4: 100% Rate of Retrofit Per Decade

Scenario 4 represents a highly optimistic case in which the entire household stock undergoes retrofitting within each decade. Under this scenario, the energy demand reduction potential is notable, with reductions ranging from 15 to 24% across minor, moderate, and major retrofit levels by 2071 (Fig. 6). These aggressive retrofit measures underscore the substantial impact of widespread, continuous energy efficiency upgrades in the residential sector. However, achieving this scenario would require overcoming significant challenges in financing, infrastructure, and policy implementation. Although idealistic, this scenario illustrates the upper-bound potential for demand reduction achievable through mass retrofitting, offering valuable insights into the scale of change possible with full policy support and streamlined processes.

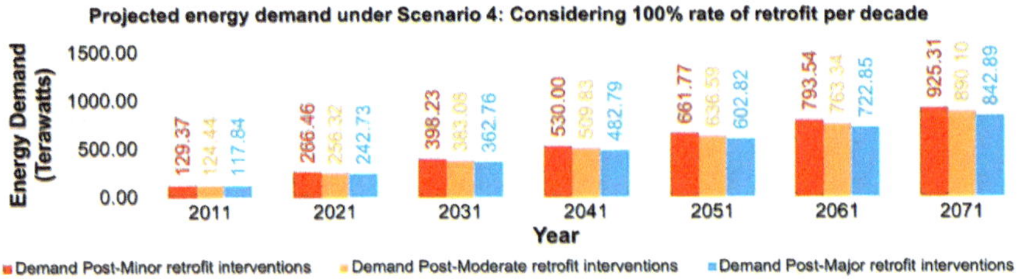

Fig. 6 Energy demand projected under Scenario 4 considering 100% rate of retrofit per decade. *Source* Author

3.2.5 Scenario 5: Phased Retrofit Policy

Scenario 5 adopts a phase-wise retrofit strategy, uniquely targeting household stock based on age and condition, starting from a 1991 base year. In this approach, buildings are categorized to undergo varying levels of retrofit: newer buildings (under 30 years) are exempt, while older structures face progressively intensive interventions from minor retrofits for moderately aged buildings to major retrofits for those over 50 years. By 2071, energy demand reductions are achieved across these targeted categories, with older and more dilapidated structures (> 80 years) showing marked improvements through intervention.

This approach reveals a reduction potential of 16.71 to 24.13% through a phased retrofit policy framework (Fig. 7). This emphasizes the importance of age-specific retrofitting for managing energy needs efficiently and underlines a more tailored method for reducing overall energy demand. Such targeted retrofitting maximizes cost-effectiveness and energy savings, emphasizing that an age-based approach can efficiently address long-term energy sustainability with tailored retrofitting levels.

The analysis of the five scenarios provides a detailed perspective on the potential energy reductions achievable through strategically implemented retrofit policies. Scenario 1 serves as the baseline, representing a status quo where no retrofit policies are applied to existing housing stock. In contrast, Scenario 4 represents an idealized

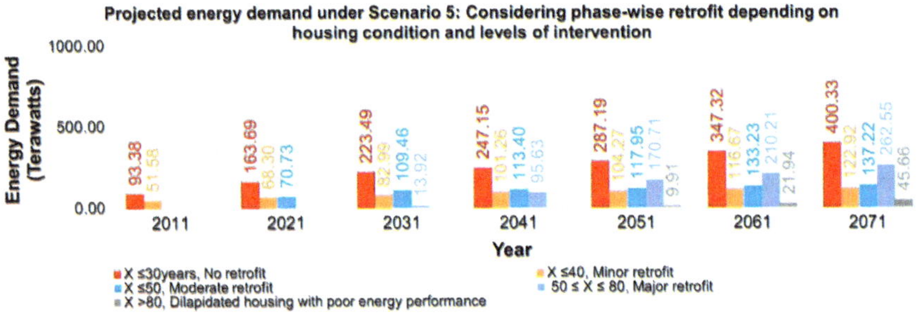

Fig. 7 Energy demand projected under Scenario 5 considering phased retrofit per decade depending on the age and condition of the building. *Source* Author

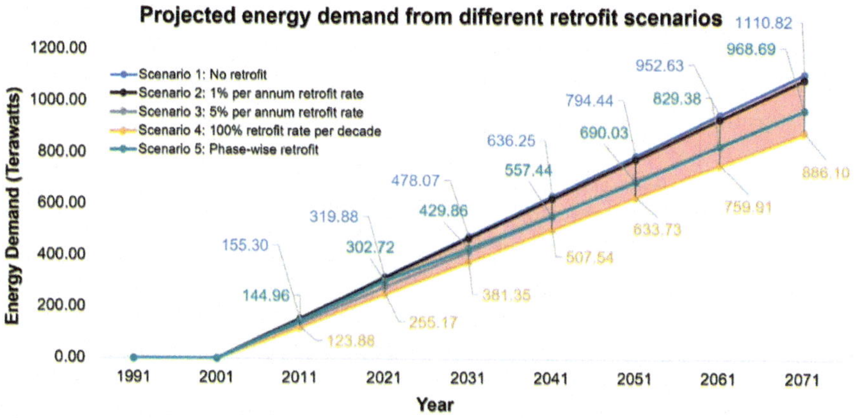

Fig. 8 Overall energy demand projected under different scenarios till 2071. *Source* Author

case, demonstrating maximum possible reductions with minimal challenges—though this level of implementation is often unrealistic in practical settings. The range of reductions that fall between these two scenarios captures the potential energy savings attainable through feasible retrofit strategies, providing insight into a realistic path for energy demand reduction (Fig. 8).

4 Conclusion

This study examined the potential and challenges of implementing energy retrofit policies in developing nations, specifically focusing on India's existing building stock. Through a hybrid methodology that combined a systematic literature review with a scenario-based analysis, the research identified significant barriers to retrofit adoption, including regulatory, economic, and socio-cultural constraints for India. The review of current energy policies and published literature highlighted the need for a comprehensive retrofit strategy tailored to the unique conditions of developing countries like India. Effective policies must focus on integrating financial incentives, awareness campaigns, and localized supply chain support to make energy retrofits feasible across diverse building sectors and income groups. Prioritizing residential and informal housing sectors, while progressively extending retrofits to public and commercial buildings, can achieve measurable reductions in energy demand.

To address the gap in quantifying the potential impact of energy retrofits, this study developed scenario-based projections of population growth and energy demand to evaluate possible savings from phased, scalable retrofit strategies. The analysis indicates that even modest annual retrofit rates, such as 1% or 5% per annum, can yield significant energy savings over time, offering a sustainable path to reduce energy demand in the residential sector. With India's commitment to achieving net-zero emissions by 2070, these incremental reductions are essential to reducing the energy production capacity required to meet this target. The scenario analysis shows that retrofit strategies could contribute to energy demand reductions of up to 25.36% in the residential sector, significantly supporting India's net-zero pathway and reducing the need to rely solely on renewable energy expansion.

The findings underscore that a phased, inclusive, and community-oriented approach to energy retrofits, supported by financial incentives and adaptable regulatory frameworks, is crucial for bridging the gap between policy aspirations and practical implementation. This study emphasizes that a strategic shift toward widespread adoption of energy retrofits can make a substantial contribution to India's energy efficiency goals, helping reduce the demand for additional energy generation and advancing towards a cleaner, more resilient energy future. By fostering zero-energy communities and building stocks, India can make meaningful progress toward its net-zero targets, positioning itself as a model for sustainable development in other emerging economies.

Appendix: Details of Energy Policies and Programs Covered Under Literature Review for Identification of Challenges and Barriers

Geographical region	Country	Income classification as per UNCTAD	Scale of implementation	Name of energy policies and programs	Source
European Union (EU)	–	–	International across EU	• Article 4 of the Energy Efficiency Directive (2012/27/EU) of the European Parliament	Bukarica and Tomsic (2017), European Union (2017), Gajić et al. (2021)
	The United Kingdom	High income	National	• The Green Deal, • The Build Upon^2 (BU^2) programs, • Together with tenants' program	Bergman and Foxon (2020), Friedler and Kumar (2019), Fylan et al. (2016), Galvin and Sunikka-Blank (2013), Gillich et al. (2018), National Housing Federation (2020)
	Germany	High income	National	• Thermal Retention Regulations (WSVO) for the New Builds,1977, • The Energy Saving Regulation (EnEV), 2002, • Drivers of Thermal Retrofit, • Tax incentive programs to support thermal retrofits in Germany, • Thermal Efficiency retrofit of Residential Buildings programs	Achtnicht and Madlener (2014), Cremer and Weber (2022), Galvin (2014), Galvin and Sunikka-Blank (2013), Weber and Wolff (2018), Wiethe and Wenninger (2023)
	Spain	High income	National	• National Building Codes, • Passivhaus Standard	Pombo et al. (2019)

(continued)

(continued)

Geographical region	Country	Income classification as per UNCTAD	Scale of implementation	Name of energy policies and programs	Source
	Slovenia	High income	National	• The Integrated National Energy and Climate Plan of the Republic of Slovenia (NECP), 2020, • Slovenian Environmental Public Fund (Eco Fund), 1993—Revolving financial fund for energy remodeling of existing buildings, • The Energy Advisory Network for households (ENSVET), 2018 • The National Energy Efficiency Action Plan (NEEAP), 2008–2016, • The Energy Act (Energetski zakon, EZ-1), 2014, • Slovenia's National Recovery and Resilience Plan (NRRP)—Target 2050	Dolšak et al. (2020), E3G (2021); E3G and SGGCCS (2021), EmBuild Project (2018), ENSVET Network (2018), European Union (2017), Gajić et al. (2021), Hrovatin and Zorić (2018), Jovović et al. (2023)

(continued)

(continued)

Geographical region	Country	Income classification as per UNCTAD	Scale of implementation	Name of energy policies and programs	Source
	Bosnia and Herzegovina	Upper middle income	National	• The Regulations on Energy efficiency in Building, 2016, • Rulebook on Minimum requirement for energy performance of buildings, • Rulebook on methodology for calculating energy performance of buildings, • Rulebook on performing energy audit of buildings and issuing RS energy certificate	Gajić et al. (2021), Kadrić et al. (2023)
	Scotland	High income	National	• The Green Deal, • Energy Company Obligation (ECO) 2013–2017, • Warm Home Discount, 2011–2021	Agkathidis and Gutiérrez (2018), Hawkey (2024), Wade et al. (2020)
	Croatia	High income	National	• The Energy Law, • The Law on Electricity Market, • The Law on Regulation of Energy activities	Gajić et al. (2021), Vrhovčak et al. (2006)
Oceania	Australia	High income	National	• Nationwide House Energy Rating Scheme (NatHERS) for existing residential properties, • Residential Efficiency Scorecard system for energy ratings	Alam et al. (2019), Bertone et al., (2016, 2018b), Department of Climate Change, Energy, the Enviroment and Water, Government of Australia (n.d.), Government of Australia (n.d.)

(continued)

(continued)

Geographical region	Country	Income classification as per UNCTAD	Scale of implementation	Name of energy policies and programs	Source
			Local area	**New South Wales:** • Energy Saving Scheme (ESS), • Government Resource Efficiency Policy (GREP) **Victoria:** • Victorian Energy Efficiency Target (VEET) • Greener Government Buildings (GGB) **South Australia:** • Retailer Energy Efficiency Scheme (REES), • Government Building Energy Strategy (GBE Strategy) **Australian Capital Territory:** • Energy-Efficient Improvement Scheme (EEIS)	
	New Zealand	High income		• New Zealand Energy Efficiency and Conservation Strategy (NZEECS), 2017–2022, • Energy efficiency requirement codes for buildings: NZS 4218:2004 and NZS 4243:2004	APEC Energy Working Group (2010), Government of New Zealand (n.d.)

(continued)

(continued)

Geographical region	Country	Income classification as per UNCTAD	Scale of implementation	Name of energy policies and programs	Source
Gulf Co-operative Council (G.C.C.)	–	–	International across G.C.C	• International Building Code or International Residential Code [IBC/IRC]	Ahmed and Asif (2021), Al-Iriani (2006), Al Naqbi et al. (2019), Alam et al. (2019), Alsahlawi (2013), Bhutto et al. (2014), Energy & Water Conservation Code (2023), Qatar (2016), Reiche (2010), Saudi Building National Code Committee (n.d.), Shamseldin (2023), UL Solutions (2023), Yang et al. (2021)
	Bahrain	High income	National	• Energy Conservation Building Code, 2018	
	Saudi Arabia	High income	National	• Building Energy Code—Vision 2030, initiated in 2007	
	Kuwait	High income	National	• Building Energy Code, • The Energy Conservation Code of Practice (R-6) for energy retrofit of buildings	
	Oman	High income	National	• Oman Energy Efficiency Labelling and Minimum Energy Performance Standard (MEPS) Scheme	
	Qatar	High income	National	• The Energy Conservation Building Code (ECBC), 2016 • Energy and Water Conservation Code, 2023	
	United Arab Emirates	High income	National	• The Dubai Demand Side Management (DSM), 2008 • The Dubai Electricity and Water Authority (DEWA), 2008 • The Green Building Regulations and Specifications in 2011	

(continued)

(continued)

Geographical region	Country	Income classification as per UNCTAD	Scale of implementation	Name of energy policies and programs	Source
Asia	Japan	High income	National	• Home Energy Management Systems-2012, • Comprehensive Assessment System for Built Environment Efficiency (CASBEE)	Bureau of Environment, Tokyo Metropolitan Authority (2018), Huang et al. (2016), Kawata and Tajima (2015)
			Local area	• The Tokyo Metropolitan Environmental Master Plan—2030 • The Tokyo Green Building Program-2002	
	Singapore	High income	National	• Singapore Green Building Master Plans—2030 and previously in (2006, 2006, 2014, 2020) • Green Mark Incentive scheme for existing buildings—2009 for financial assistance	B.C.A, S.G.B.C (2021), Hwang et al. (2017), Lim (2021)
			Local area	• The Housing Development Board (HDB) Green print at Yuhua Estate—2012 to 2015	
	China	Upper middle income	National	• The Clean Development Mechanism (CDM)—Revolving financial fund policy • National Energy Conservation Publicity Week (1991–2022)	Huang et al. (2016), Liu et al. (2020), Liu et al. (2023), UNFCCC (2023), Zhou et al. (2022)

(continued)

(continued)

Geographical region	Country	Income classification as per UNCTAD	Scale of implementation	Name of energy policies and programs	Source
			Local area	• Local area plans for each provincial government • Subsidy plans "Rewards in Lieu method" • Mandatory Energy Building Codes with incentive programs for existing buildings in Beijing	
	South Korea	High income	National	• The Green Remodelling Energy Simulation Program (GR-E), 2020 • The Framework Act on Low Carbon Green Growth, 2010 • Korean green building certification system (K-GBCS), 2002 • The Energy Act: September 2006 • Green Remodelling Interest Subsidy Business (GRISB), • Green New Deal (GND), 2021 • Korean Building Energy Standard, 2004	An et al. (2023), Asia Business Council (2020), Kim and Lim (2021), Lee et al. (2016), M.O.T.I.E. (2019), Nachmany et al. (2015), Painuly et al. (2003)
America and Canada	North America	High income	National	• The Clean Energy Rule, under Section 433(a) of the Energy Independence and Security Act, 2007, tending to public and non-public federal buildings	D'Oca et al. (2018), Grueneich et al. (n.d.), Martinez et al. (2010). U.S. Department of Commerce (1998), Zhang et al. (2021)

(continued)

(continued)

Geographical region	Country	Income classification as per UNCTAD	Scale of implementation	Name of energy policies and programs	Source
	Canada	High income	National	• National Energy Code of Canada for Buildings (NECB), 2020	Government of Canada (n.d.)

References

Achtnicht, M., & Madlener, R. (2014). Factors influencing German house owners' preferences on energy retrofits. *Energy Policy, 68*, 254–263. https://doi.org/10.1016/j.enpol.2014.01.006

Agkathidis, A., & Gutiérrez, R. U. (2018). *Sustainable retrofits: Post-War residential towers in Britain (First).* Routledge.

Ahmed, W., & Asif, M. (2021). A critical review of energy retrofitting trends in residential buildings with particular focus on the GCC countries. *Renewable and Sustainable Energy Reviews, 144*(June 2020), 111000. https://doi.org/10.1016/j.rser.2021.111000

Al-Iriani, M. A. (2006). Energy-GDP relationship revisited: An example from GCC countries using panel causality. *Energy Policy, 34*(17), 3342–3350. https://doi.org/10.1016/j.enpol.2005.07.005

Al Naqbi, S., Tsai, I., & Mezher, T. (2019). Market design for successful implementation of UAE 2050 energy strategy. *Renewable and Sustainable Energy Reviews, 116*(October), 109429. https://doi.org/10.1016/j.rser.2019.109429

Ala-Juusela, M., & ur Rehman, H. (2020). PEB as enabler for consumer centred clean energy transition-shared definition and concept; EXCESS Deliverable 1.1. *Excess, Proposal No. 870157*. https://positive-energy-buildings.eu/resource?t=ReportonthedefinitionofaPositiveEnergyBuilding (PEB)

Alam, M., Zou, P. X. W., Stewart, R. A., Bertone, E., Sahin, O., Buntine, C., & Marshall, C. (2019). Government championed strategies to overcome the barriers to public building energy efficiency retrofit projects. *Sustainable Cities and Society, 44*(June 2018), 56–69. https://doi.org/10.1016/j.scs.2018.09.022

Alsahlawi, M. A. (2013). Measuring Energy Efficiency in GCC countries using data envelopment analysis. *Journal of Business Inquiry: Research, 12*(1), 15–30.

An, J., Jung, D., Jeong, K., Ji, C., Hong, T., Lee, J., Kapp, S., & Choi, J. (2023). Energy-environmental-economic assessment of green retrofit policy to achieve 2050 carbon-neutrality in South Korea: Focused on residential buildings. *Energy and Buildings, 289*, 113059. https://doi.org/10.1016/j.enbuild.2023.113059

APEC Energy Working Group. (2010). *Building Energy Codes Report —Singapore* (Issue January). https://www.apec.org/docs/default-source/Publications/2010/1/Building-Energy-Codes-Report-for-New-Zealand-January-2010/210_ewg_EnergyCodesNZRpt.pdf

Asia Business Council. (2020). *Energy efficiency building standards in South Korea.* http://www.asiabusinesscouncil.org/docs/BEE/papers/BEE_Policy_Korea.pdf

B.C.A, S.G.B.C. (2021). *Singapore Green Building Masterplan.* https://www1.bca.gov.sg/docs/default-source/docs-corp-buildsg/sustainability/20220726_singapore-green-building-masterplan-booklet.pdf

Bergman, N., & Foxon, T. J. (2020). Reframing policy for the energy efficiency challenge: Insights from housing retrofits in the United Kingdom. *Energy Research and Social Science, 63*(November 2018), 101386. https://doi.org/10.1016/j.erss.2019.101386

Bertone, E., Sahin, O., Stewart, R. A., Zou, P., Alam, M., & Blair, E. (2016). State-of-the-art review revealing a roadmap for public building water and energy efficiency retrofit projects. *International Journal of Sustainable Built Environment, 5*(2), 526–548. https://doi.org/10.1016/j.ijsbe.2016.09.004

Bertone, E., Stewart, R. A., Sahin, O., Alam, M., Zou, P. X. W., Buntine, C., & Marshall, C. (2018). Guidelines, barriers and strategies for energy and water retrofits of public buildings. *Journal of Cleaner Production, 174*(November), 1064–1078. https://doi.org/10.1016/j.jclepro.2017.11.065

Berwal, A. K., & Yadav, M. (2021). Retrofit strategy in existing building for implementing energy conservation building code in India. *European Journal of Engineering and Technology Research, 6*(5), 134–140. https://doi.org/10.24018/ejers.2021.6.5.2535

Bhutto, A. W., Bazmi, A. A., Zahedi, G., & Klemeš, J. J. (2014). A review of progress in renewable energy implementation in the Gulf Cooperation Council countries. *Journal of Cleaner Production, 71*(2014), 168–180. https://doi.org/10.1016/j.jclepro.2013.12.073

Brown, M. (2014). Innovative energy-efficiency policies: an international review. *WIREs Energy and Environment*, *4*(1), 1–25. https://doi.org/10.1002/wene.125

Bukarica, V., & Tomsic, Z. (2017). Design and evaluation of policy instruments for energy efficiency market. *IEEE Transactions on Sustainable Energy, 8*(1), 354–362. https://doi.org/10.1109/TSTE.2016.2599584

Bureau of Environment, Tokyo Metropolitan Authority. (2018). *Green building program*. https://www.kankyo.metro.tokyo.lg.jp/en/climate/build.html#:~:text=Thesystemwasintroducedinsubmittheirbuildingenvironmentplans

Census of India. (2011). *Distribution of households by the condition of census houses occupied by them (all)*. Office of the Registrar General & Census Commissioner, India. Distribution of households by the condition of census houses occupied by them (all).

Chaturvedi, V., Eom, J., Clarke, L. E., & Shukla, P. R. (2014). Long term building energy demand for India: Disaggregating end use energy services in an integrated assessment modeling framework. *Energy Policy, 64*, 226–242. https://doi.org/10.1016/j.enpol.2012.11.021

Cremer, L., & Weber, C. (2022). Deep energy retrofits: How effective and robust are policy instruments? *Energy Policy*, *170*(August 2021). https://doi.org/10.1016/j.enpol.2022.113210

D'Oca, S., Ferrante, A., Ferrer, C., Pernetti, R., Gralka, A., Sebastian, R., & Veld, P. op't. (2018). Technical, financial, and social barriers and challenges in deep building renovation: Integration of lessons learned from the H2020 cluster projects. *Buildings*, *8*(12). https://doi.org/10.3390/buildings8120174

Dandia, G., Sudhakaran, P., & Basu, C. (2021). Evaluation of energy-efficient retrofit potential for government offices in India. *Architecture and Engineering*, *6*(2), 03–17. https://doi.org/10.23968/2500-0055-2021-6-2-03-17

Department of Climate Change, Energy, the Enviroment and Water, Government of Australia. (n.d.). *Supporting energy efficiency upgrades for existing homes through informed policy and program design: Finance Industry perspectives*. https://www.dcceew.gov.au/sites/default/files/documents/supporting-energy-efficiency-upgrades-for-existing-homes-through-informed-policy-and-program-design-finance-industry-perspectives.pdf

Dey, P., & Keshri, S. (2023). Retrofitting: Exploring a new domain in energy-efficient built environment. In J. Vijayalaxmi (Ed.), *Building construction and technology* (1st ed., pp. 175–188). Springer. https://doi.org/10.1007/978-981-99-3526-0_12

Dolšak, J., Hrovatin, N., & Zorić, J. (2020). Factors impacting energy-efficient retrofits in the residential sector: The effectiveness of the Slovenian subsidy program. *Energy and Buildings*, *229*. https://doi.org/10.1016/j.enbuild.2020.110501

E3G. (2021). Renovate 2 recover: How transformational are the national recovery plans for buildings renovation? In *REDay2021 Conference*. https://www.renovate-europe.eu/wp-content/uploads/2018/09/Renovate2Recover_Full-Study-1.pdf

E3G, & SGGCCS. (2021). Renovate2Recover—Slovenia. In *Renovate2Recover: How transformational are the national recovery plans for buildings renovation?* http://www.renovate-europe.eu/wp-content/uploads/2018/09/Renovate2Recover-SLOVENIA.pdf

EmBuild Project. (2018). *A snapshot of building renovation strategies*. 28. http://bpie.eu/publication/a-snapshot-of-building-renovation-strategies-bulgaria-croatia-germany-romania-slovenia/

ENSVET Network. (2018). *Build2LC—Energy Consultancy*. https://projects2014-2020.interregeurope.eu/fileadmin/user_upload/tx_tevprojects/library/file_1540905584.pdf

European Union. (2017). National Energy Efficiency Action Plan. In *National energy efficiency action plan* (Issue December). https://energy.ec.europa.eu/system/files/2018-04/si_neeap_2017_en_0.pdf

Friedler, C., & Kumar, C. (2019). *Reinventing retrofit: How to scale up home energy efficiency in the UK*. 1–19.

Fylan, F., Glew, D., Smith, M., Johnston, D., Brooke-Peat, M., Miles-Shenton, D., Fletcher, M., Aloise-Young, P., & Gorse, C. (2016). Reflections on retrofits: Overcoming barriers to energy efficiency among the fuel poor in the United Kingdom. *Energy Research and Social Science, 21*, 190–198. https://doi.org/10.1016/j.erss.2016.08.002

Gajić, D., Peulić, S., Mavrič, T., Sandak, A., Tavzes, Č, Malešević, M., & Slijepčević, M. (2021). Energy retrofitting opportunities using renewable materials—Comparative analysis of the current frameworks in Bosnia-Herzegovina and Slovenia. *Sustainability (Switzerland), 13*(2), 1–19. https://doi.org/10.3390/su13020603

Galvin, R. (2014). Why German homeowners are reluctant to retrofit. *Building Research and Information, 42*(4), 398–408. https://doi.org/10.1080/09613218.2014.882738

Galvin, R., & Sunikka-Blank, M. (2013). Economic viability in thermal retrofit policies: Learning from ten years of experience in Germany. *Energy Policy, 54*, 343–351. https://doi.org/10.1016/j.enpol.2012.11.044

Gillich, A., Sunikka-Blank, M., & Ford, A. (2018). Designing an 'optimal' domestic retrofit programme. *Building Research and Information, 46*(7), 767–778. https://doi.org/10.1080/09613218.2017.1368235

Government of Australia. (n.d.). *Expanding NatHERS assessments Ratings for existing homes*. Retrieved March 2, 2024, from https://www.nathers.gov.au/expansion

Government of Canada. (n.d.). *Canada's national energy code*. Retrieved September 25, 2024, from https://natural-resources.canada.ca/energy-efficiency/buildings/new-buildings/canadas-national-energy-code/20675

Government of New Zealand. (n.d.). *H1 energy efficiency*. Ministry of Business, Innovation & Employment. Retrieved September 24, 2024, from https://www.building.govt.nz/building-code-compliance/h-energy-efficiency/h1-energy-efficiency

Grueneich, D. M., Advocates, G. R., London, J. S., & Advocates, G. R. (n.d.). *Update on California's approach to energy efficiency status of investor-owned utility energy-efficiency programs in California* (pp. 116–128).

Hawkey, D. (2024). *No home left behind funding a just transition to clean heat in Scotland. February*.

Hrovatin, N., & Zorić, J. (2018). Determinants of energy-efficient home retrofits in Slovenia: The role of information sources. *Energy and Buildings, 180*(December 2009), 42–50. https://doi.org/10.1016/j.enbuild.2018.09.029

Huang, B., Mauerhofer, V., & Geng, Y. (2016). Analysis of existing building energy saving policies in Japan and China. *Journal of Cleaner Production, 112*, 1510–1518. https://doi.org/10.1016/j.jclepro.2015.07.041

Hwang, B. G., Shan, M., Xie, S., & Chi, S. (2017). Investigating residents' perceptions of green retrofit program in mature residential estatesThe case of Singapore. *Habitat International, 63*, 103–112. https://doi.org/10.1016/j.habitatint.2017.03.015

International Energy Agency. (2019). Final energy consumption and efficiency. *World Energy Outlook*. https://www.iea.org/weo, https://www.iea.org/reports/world-energy-outlook-2019

International Energy Agency. (2021). *Pathway to critical and formidable goal of net-zero emissions by 2050 is narrow but brings huge benefits, according to IEA special report*. https://www.iea.org/news/pathway-to-critical-and-formidable-goal-of-net-zero-emissions-by-2050-is-narrow-but-brings-huge-benefits

International Energy Agency. (2022a). *Buildings*. https://www.iea.org/reports/buildings

International Energy Agency. (2022b). *For the first time in decades, the number of people without access to electricity is set to increase in 2022*. https://www.iea.org/commentaries/for-the-first-time-in-decades-the-number-of-people-without-access-to-electricity-is-set-to-increase-in-2022

Jovović, I., Cirman, A., Hrovatin, N., & Zorić, J. (2023). Do social capital and housing-related lifestyle foster energy-efficient retrofits? Retrospective panel data evidence from Slovenia. *Energy Policy, 179*(June 2022). https://doi.org/10.1016/j.enpol.2023.113651

Kadrić, D., Aganović, A., & Kadrić, E. (2023). Multi-objective optimization of energy-efficient retrofitting strategies for single-family residential homes: Minimizing energy consumption, CO_2 emissions and retrofit costs. *Energy Reports, 10*, 1968–1981. https://doi.org/10.1016/j.egyr.2023.08.086

Kawata, K., & Tajima, M. (2015). Evaluation of subsidy systems for energy retrofit of existing dwellings in Japan. *Internet Journal for Society for Social Management Systems, 10*(1). https://core.ac.uk/download/pdf/75848508.pdf

Kim, J., & Lim, S. (2021). A direction to improve EER (Energy Efficiency Retrofit) policy for residential buildings in South Korea by means of the recurrent EER policy. *Sustainable Cities and Society, 72*(February), 103049. https://doi.org/10.1016/j.scs.2021.103049

Kumbaroglu, G., & Madlener, R. (2012). Evaluation of economically optimal retrofit investment options for energy savings in buildings. *Energy and Buildings, 49*, 327–334. https://doi.org/10.1016/j.enbuild.2012.02.022

Lee, K., Koo, B., Park, B., & Ahn, Y. H. (2016). The development of an energy-efficient remodeling framework in South Korea. *Habitat International, 53*, 430–441. https://doi.org/10.1016/j.habitatint.2015.12.003

Lim, L. Y. (2021). *Singapore green building masterplan public engagement report* (Issue February). https://iea.blob.core.windows.net/assets/9d80b582-fe7b-46c3-9fde-0f85e016116f/SGBMPsharingforIEAwebinar20May2021.pdf

Liu, G., Tan, Y., & Li, X. (2020). China's policies of building green retrofit: A state-of-the-art overview. *Building and Environment, 169*(November 2019), 106554. https://doi.org/10.1016/j.buildenv.2019.106554

Liu, Z., Yu, C., Qian, Q. K., Huang, R., You, K., Visscher, H., & Zhang, G. (2023). Incentive initiatives on energy-efficient renovation of existing buildings towards carbon–neutral blueprints in China: Advancements, challenges and prospects. *Energy and Buildings, 296*(June). https://doi.org/10.1016/j.enbuild.2023.113343

M.O.T.I.E. (2019). *A new energy paradigm for the future: Third energy master plan.* Ministry of Trade, Industry and Energy; South Korea. https://www.etrans.or.kr/ebook/05/files/assets/common/downloads/ThirdEnergyMasterPlan.pdf

Martinez, S., Wang, D., & Chou, J. (2010). *California restores its energy efficiency leadership—Smart policies provide enormous economic and environmental benefits.* Natural Resources Defense Council.

Menassa, C. C. (2011). Evaluating sustainable retrofits in existing buildings under uncertainty. *Energy and Buildings, 43*(12), 3576–3583. https://doi.org/10.1016/j.enbuild.2011.09.030

MOSPI, G. of I. (2021). *Average household size in India, 2020–21.* Ministry of Statistics and Programme Implementation. https://www.mospi.gov.in/estimated-number-households-average-household-size-and-sex-ratio-no-female-1000-male-2

Nachmany, M., Fankhauser, S., Davidová, J., Kingsmill, N., Landesman, T., Roppongi, H., Schleifer, P., Setzer, J., Sharman, A., Singleton, C. S., Sundaresan, J., & Townshend, T. (2015). *Climate change legislation in South Korea.* https://www.lse.ac.uk/GranthamInstitute/wp-content/uploads/2015/05/SOUTH_KOREA.pdf

National Housing Federation. (2020). *The Together with Tenants charter aims to strengthen the relationship between residents and housing association landlords.* https://www.housing.org.uk/our-work/together-with-tenants/

van Eck, N. J., & Waltman, L. (2010). *VOSviewer* (1.6.19). Centre for Science and Technology Studies (CWTS). https://www.vosviewer.com/

NITI Aayog, Government of India. (2019). *India energy dashboard—Electricity consumption.* https://www.niti.gov.in/edm/#elecConsumption

Painuly, J. P., Park, H., Lee, M. K., & Noh, J. (2003). Promoting energy efficiency financing and ESCOs in developing countries: Mechanisms and barriers. *Journal of Cleaner Production, 11*(6), 659–665. https://doi.org/10.1016/S0959-6526(02)00111-7

Pombo, O., Rivela, B., & Neila, J. (2019). Life cycle thinking toward sustainable development policy-making: The case of energy retrofits. *Journal of Cleaner Production, 206*, 267–281. https://doi.org/10.1016/j.jclepro.2018.09.173

Pyloudi, E., Papantoniou, S., & Kolokotsa, D. (2015). Retrofitting an office building towards a net zero energy building. *Advances in Building Energy Research, 9*(1), 20–33. https://doi.org/10.1080/17512549.2014.917985

Energy & Water Conservation Code 2023, Qatar. (2016). https://www.km.qa/CustomerService/ServiceRegulations/ENERGYandWATERCONSERVATIONCODE2023.pdf

Reiche, D. (2010). Energy Policies of Gulf Cooperation Council (GCC) countries-possibilities and limitations of ecological modernization in rentier states. *Energy Policy, 38*(5), 2395–2403. https://doi.org/10.1016/j.enpol.2009.12.031

Saudi Building National Code Commitee. (n.d.). *The SBC's energy conservation requirements.* https://sbc.gov.sa/En/Pages/default.aspx

Shamseldin, A. (2023). Proposed role of the local Saudi building codes in assessing the energy performance of buildings in KSA's GBRS. *Ain Shams Engineering Journal, 14*(5), 101966. https://doi.org/10.1016/j.asej.2022.101966

Sharma, S. K., Mohapatra, S., Sharma, R. C., Alturjman, S., Altrjman, C., Mostarda, L., & Stephan, T. (2022). Retrofitting existing buildings to improve energy performance. *Sustainability (Switzerland), 14*(2), 1–14. https://doi.org/10.3390/su14020666

Surya Prakash, B., Archana, S., Aravind, & Priya Dharshini, E. (2021). Adoption of retrofitting catalogue for cutting down electricity consumption of residences in warm and humid region. *XIII*(4), 397–419. https://www.xajzkjdx.cn/gallery/38-april2021.pdf

T.E.R.I. (2013). *Roadmap for incorporating energy efficiency retrofits in existing buildings.* The Energy and Resources Institute Press. http://cbs.teriin.org/pdf/Energy_Efficiency_Retrofits_in_Existing_Buildings.pdf

The United Nations Climate Change Conference. (2021). *Conference of Parties 26.* https://www.un.org/en/climatechange/cop26

U.S. Department of Commerce. (1998). *BEES—LCA Guideline.pdf*. https://www.nist.gov/publications/bees-10-building-environmental-and-economic-sustainability-technical-manual-and-user

UL Solutions. (2023). *Bahrain and Oman energy efficiency labelling and MEPS scheme* (Issue December, pp. 2–5). UL Solutions. https://www.ul.com/sites/default/files/2023-12/Bahrain-And-Oman-EE-Service-Terms-30Nov2032.pdf

UNCTAD. (n.d.). *Country classification—Geographical groups.* Retrieved September 24, 2024, from https://unctadstat.unctad.org/EN/Classifications.html

UNFCCC. (2016). *The Paris agreement.* What is the Paris agreement? https://unfccc.int/process-and-meetings/the-paris-agreement

UNFCCC. (2023). *The China clean development mechanism fund (CCDMF).* Financing for Climate Friendly Investment, The United Nations Framework Convention on Climate Change. https://unfccc.int/climate-action/momentum-for-change/financing-for-climate-friendly/china-clean-development-mechanism-fund

Vrhovčak, M. B., Tomšić, Ž., & Debrecin, N. (2006). Potential and use of renewable energy sources in Croatia. *Renewable Energy, 31*(12), 1867–1872. https://doi.org/10.1016/j.renene.2005.09.021

Wade, F., Bush, R., & Webb, J. (2020). Emerging linked ecologies for a national scale retrofitting programme: The role of local authorities and delivery partners. *Energy Policy, 137*(December 2019), 111179. https://doi.org/10.1016/j.enpol.2019.111179

Weber, I., & Wolff, A. (2018). Energy efficiency retrofits in the residential sector—Analysing tenants' cost burden in a German field study. *Energy Policy, 122*(August), 680–688. https://doi.org/10.1016/j.enpol.2018.08.007

Wiethe, C., & Wenninger, S. (2023). The influence of building energy performance prediction accuracy on retrofit rates. *Energy Policy, 177*(April 2022), 113542. https://doi.org/10.1016/j.enpol.2023.113542

Yang, B., Jahanger, A., Usman, M., & Khan, M. A. (2021). The dynamic linkage between globalization, financial development, energy utilization, and environmental sustainability in GCC countries. *Environmental Science and Pollution Research, 28*(13), 16568–16588. https://doi.org/10.1007/s11356-020-11576-4

Zhang, H., Hewage, K., Karunathilake, H., Feng, H., & Sadiq, R. (2021). Research on policy strategies for implementing energy retrofits in the residential buildings. *Journal of Building Engineering, 43*(August), 103161. https://doi.org/10.1016/j.jobe.2021.103161

Zhou, H., Zhao, Y., Zhang, Z., Geng, Y., Yu, J., & Lin, B. (2022). Post occupancy investigation of 40 certified green buildings in Beijing: Results, lessons and policy suggestions. *Journal of Building Engineering, 60*(June), 105153. https://doi.org/10.1016/j.jobe.2022.105153

Indoor Thermal Performance of Roof Using Vernacular Material in Hot Semi-arid Region

Chandu Sai Tarun and J. Vijayalaxmi

Abstract Building materials are the primary factors along with socio-cultural structure, climate, technology and economy that influence architectural forms. Vernacular construction techniques using Kadapa stone are used in the hot semi-arid region of Kadapa, India. However, no quantitative or qualitative evaluation has been used to establish its thermal performance. This study compares the indoor environmental conditions of vernacular dwellings with kadapa-lime roof and a modern house with RCC roof using qualitative and quantitative analyses. Both the houses have walls made of 0.6 M thick Kadapa stone. Field measurements of indoor air temperature and humidity are analysed and compared based on adaptive thermal comfort model in the vernacular house and a modern house simultaneously. A post-occupancy evaluation is performed using a questionnaire survey to compare the outcomes of the field study's findings. Results of the study are plotted based on adaptive thermal comfort model with an indoor operative temperature range of 28.95–33.71 °C. Temperatures inside the vernacular house are the closest to the comfort range with a deviation of only 0.1–1 °C above the upper comfort limit whereas the temperatures inside the modern house have a deviation of 1–3.6 °C above the upper comfort limit. All the occupants in modern houses voted for uncomfortable or very uncomfortable conditions and 57% of the occupants from vernacular houses voted for slightly comfortable and comfortable conditions. It is also found that RCC roof slabs are not an appropriate choice for roof construction in hot semi-arid conditions as it did not perform well thermally as compared to the kadapa-lime roof. Vernacular buildings with local materials like Kadapa stone, lime concrete roof and other passive strategies helped in achieving close to comfort conditions during the summer time in the hot semi-arid climate.

Keywords Climate responsiveness · Indoor thermal comfort · Kadapa · Vernacular architecture

---

C. S. Tarun (✉)
Sustainable Architecture, School of Planning and Architecture, Vijayawada, India
e-mail: chandusaitarun@gmail.com

J. Vijayalaxmi
School of Planning and Architecture, Vijayawada, India

J. Vijayalaxmi (ed.), *Sustainable Building Performance*,
https://doi.org/10.1007/978-981-96-2428-7_3

1 Introduction

Vernacular architecture is a concept that describes construction practices that make use of locally accessible materials in response to local climate and geography. Vernacular architecture develops over a period of time to reflect the environmental, cultural, and historical circumstances of the time period. It refers to architecture that is specific to a period or location. (Vijayalaxmi & Arathy, 2022). Vernacular housing evolved over the years as a response to climate, local materials, geography, socio-cultural practices and economic and occupational needs. Architects and engineers should look for answers in vernacular structures for designing low-energy, environmentally friendly and localised entities while using current materials and techniques, for a sustainable development (Dili et al., 2010). A thorough understanding of the materials and construction practices used and strategies adopted will be valuable in current construction if adopted wisely. The need of the hour is to provide sustainable and energy-efficient housing solutions.

Human thermal comfort varies depending on climatic zones and environmental factors. About 45% of energy utilised in residential buildings in India is used to address thermal comfort needs. (ECBC, 2007). Air conditioners in residences are being used to achieve thermal comfort and are a major contributor to rising energy consumption (Eco-Niwas Samhitha, 2018; Vijayalaxmi & Sekar, 2013).

Previous research on indoor thermal comfort in India concentrated on midrise apartments in Hyderabad, traditional houses in northeastern India, Kerala, and Tamil Nadu, as well as student hostels and office typology buildings in composite climates (Kumar et al., 2016). Thermal comfort studies in Indian residential buildings: a review based on science mapping approach (Fig. 1), shows that no such study was done in the state of Andhra Pradesh, especially in hot semi-arid climatic conditions (Chellappa & Srivastava, 2022). This paper aims to understand the material-dictated construction practices adopted in vernacular architecture and assess its indoor comfort in a hot semi-arid climate. This paper aims to investigate one such vernacular housing for its indoor comfort that has been primarily dependent on a single material-based construction.

Joshima et al. (2021) conducted field studies on five different types of roof coverings on RCC slabs namely Mangalore tile, ceramic tile, exposed RCC, concrete tile and shingles. It is concluded that the exposed RCC slab has the worst thermal performance.

The main goals of this research are to: (1) Identify construction practices and techniques which developed as a result of the response to climate and resource availability in vernacular buildings, (2) assess the indoor thermal performance and thermal comfort thus achieved as result of the local construction materials and practices.

This paper presents a detailed examination of construction techniques and practices in the vernacular architecture of the hot semi-arid region of Kadapa based on a field investigation with an emphasis on indoor thermal comfort.

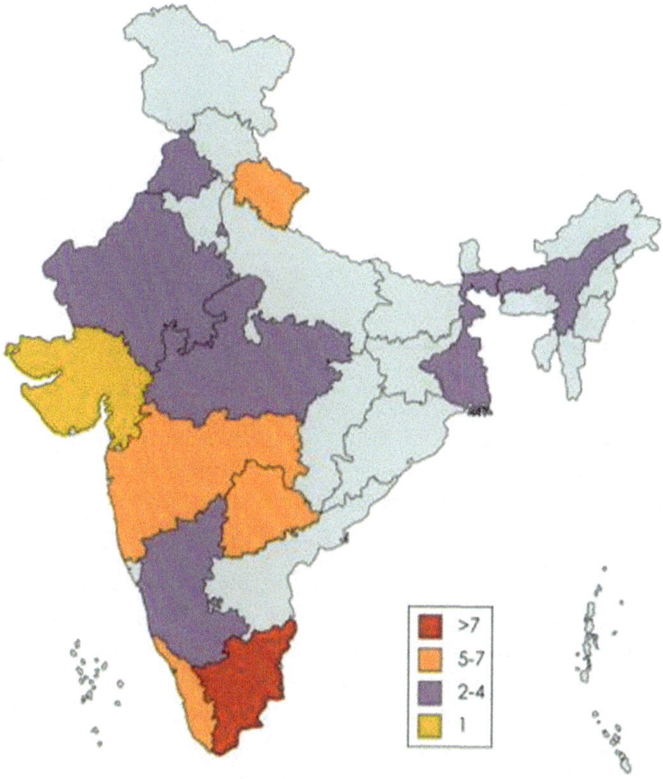

Fig. 1 Geographical distribution of thermal comfort in residential building studies in India (2001–2020). *Source* (Chellappa & Srivastava, 2022).

2 Study Area

2.1 Geographical Location

The area of study is in Andhra Pradesh's Kadapa region (Fig. 2). Jammalamadugu (14.83° N 78.4° E). It is 169 m above the mean sea level. This town is famously known for the mining of the local black kadapa limestone which is being used in construction till date.

2.2 Climate

Kadapa is classified under BSh as per Köppen climate classification and is a semi-arid region. It experiences high temperatures throughout the year. Temperatures reach up

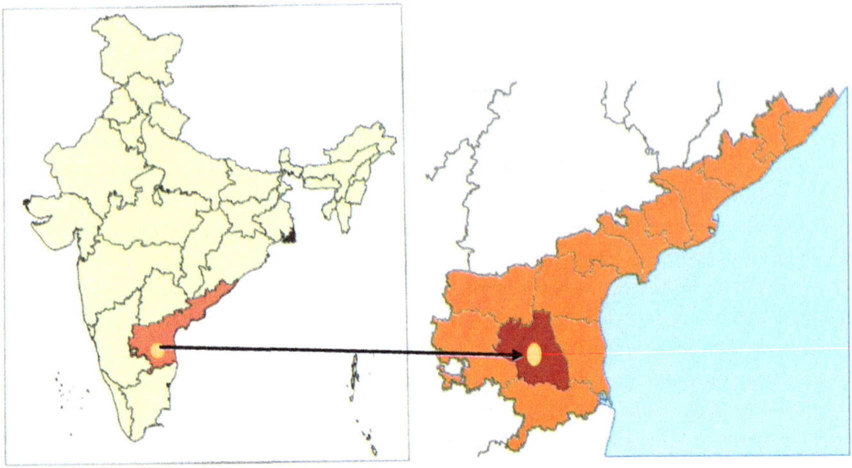

Fig. 2 Study area—Jammalamadugu, Kadapa

to 46 °C. Summer months are very uncomfortable because of the hot and humid weather. Temperatures range varies from a low of 26 °C to a high of 46 °C during summers. The average relative humidity during the months of March–May is under 35% and above 50% during the monsoon months.

3 Vernacular Buildings in Kadapa Region: An Overview

3.1 Form and Planning

The vernacular houses of the Kadapa region are row houses with shared walls. The houses are long and narrow. Most of the houses have a longer axis along the North–South orientation with shared walls on the East and West. The houses open onto the streets in either North or Southern directions with a courtyard on the other side. The houses are typically row houses with the longer axis along North–South directions as shown in Fig. 3b. The walls on the East and West directions which have a high direct sun radiation, are shared walls and hence are devoid of heat ingress. A typical vernacular house welcomes with a shaded verandah also called an *arugu* for guests and informal activities (Fig. 3a). It opens into a large hall with centrally placed small skylight. The hall is followed by additional rooms and a kitchen. Courtyards are provided on the northern side which also constitute wet areas of the house. Small windows are used mostly in the North and South directions. Doors are aligned linearly to facilitate cross ventilation. With shared walls not exposed to direct sunlight, roof is the major contributor to heat gain in these houses.

a) b)

Fig. 3 Vernacular row housing **a** front view **b** planning and orientation

3.2 Construction Materials and Techniques

Walls

Traditionally 600 mm thick load-bearing stone masonry walls are constructed with kadapa stone with lime for mortar and plaster. The hygroscopic nature of the lime plaster helps in regulating humidity in the air. The walls on the East and West direction of these houses are mostly shared walls. Both the vernacular house and the modern house chosen for the study have the same walling material made of 0.6 M thick Kadapa stone but with cement plaster. Coming from the limestone family, the 600 mm thick stone walls have a high thermal capacity and can delay the heat gain in the indoors until late in the night as they can have higher time lag. Traditionally the people of this region sleep in the courtyards or on the terraces during the night when the building is cooled by night flushing.

Roofing

The vernacular roofs are constructed with lime concrete. Wooden logs placed at 0.3 M c/c are utilised as spanning components and as the roof's foundation. These are supported by stone projections on the wall. Kadapa stone slabs are laid over an array of wooden rafters (Fig. 3). Sunnam (lime) mortar is used to fill in the gaps between the large stone slabs. Over this, a mixture of kadapa stone aggregate and sand is laid. Chalk lime stones are laid on top and water is added. The lime reacts and is manually mixed thoroughly to form the concrete. The final finish is a nearly 50 mm thick layer of lime. When this hardens, the flat lime roof becomes waterproof and leak-proof.

Coming from the limestone family, the thick stone walls and roof have a high thermal capacity and can delay the heat gain in the indoors until late in the night as they can have higher time lag. Traditionally the people of this region sleep in the courtyards or on the terraces during the night when the building is cooled by night flushing. The roof surfaces lose the heat to the night sky much quicker.

4 SDGs Addressed by Sustainable Architecture

Sustainable architecture plays a crucial role in addressing multiple SDGs, particularly SDG 3 (Good Health and Well-being), SDG 7 (Affordable and Clean Energy), SDG 9 (Industry, Innovation, and Infrastructure), and SDG 13 (Climate Action) (Fig. 4). This research on vernacular architecture in India, particularly in the hot semi-arid region of Kadapa, contributes to addressing the above-mentioned SDGs through its focus on traditional building practices, energy efficiency, and indoor environmental quality. The study, which compares the thermal performance of vernacular dwellings with modern constructions with a focus on encapsulating the effect of roof, highlights how local materials and climate-responsive design strategies can significantly improve thermal comfort and reduce energy consumption.

SDG 3: Good Health and Well-being: The paper demonstrates how vernacular architecture, through its use of Kadapa stone, lime concrete for roof provides a healthier and more comfortable indoor environment compared to modern construction with reinforced cement concrete (RCC) roofs, which leads to higher internal temperatures.

SDG 7: Affordable and Clean Energy: The kadapa-lime roof effectively mitigates heat transfer, keeping indoor temperatures within or near comfort ranges. In contrast, modern houses with RCC roofs show higher internal temperatures, leading to discomfort and greater energy use for cooling. By showcasing these energy-saving strategies in vernacular architecture, the paper aligns with SDG 7 by promoting affordable and energy-efficient housing solutions that reduce the need for artificial cooling and decrease energy demand.

SDG 9: Industry, Innovation, and Infrastructure: The paper underscores the importance of locally-sourced materials and traditional building techniques, which reduce the environmental impact of construction and foster innovation in sustainable

Fig. 4 SDGs addressed by sustainable architecture

design. The use of Kadapa stone, a locally abundant material, minimises the energy required for transportation and processing, contributing to a lower carbon footprint for construction projects. The techniques and materials used can be adapted to modern infrastructure needs, creating resilient and climate-responsive buildings.

SDG 13: Climate Action: By promoting these climate-appropriate building methods, the study supports climate adaptation and mitigation strategies in India's construction sector. The insights gained from field measurements and the adaptive thermal comfort model can help inform policies and building codes, encouraging the adoption of low-energy, climate-resilient construction practices.

5 Methodology

Field studies have been carried out to document two houses with vernacular Kadapa stone walls but roofing with Kadapa stone-lime concrete roof and modern RCC slab. The construction practices in both the houses were documented and analysed. A vernacular row house and a modern house, both naturally ventilated were identified for the study. A field study is done to measure the indoor air temperature and relative humidity in both the houses simultaneously for a duration of 24 h. The workflow and the methodology followed as shown in Fig. 5.

A post-occupancy evaluation is performed using a questionnaire survey to compare the results of the findings from the field study (Dili et al., 2010). The

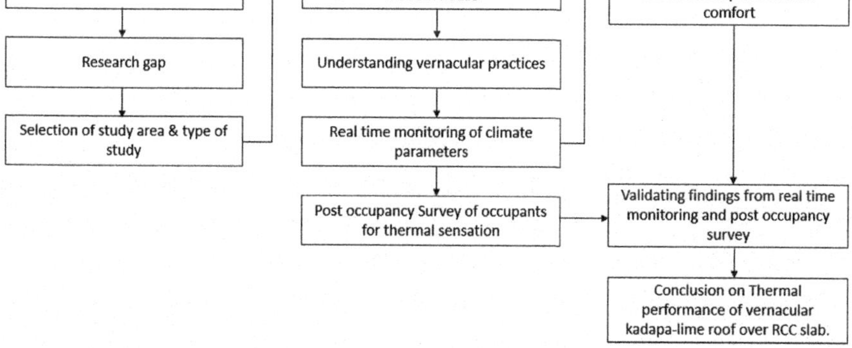

Fig. 5 Research process and methodology

questionnaire survey is used to note the occupants' sensation with respect to indoor temperature, humidity, and overall thermal comfort. Occupants from both vernacular and modern houses within the same village are surveyed after informed consent (NBC, 2016).

5.1 Adaptive Thermal Comfort Model

Comfort limits for temperature indoors are calculated using the neutral temperature which is the indoor operative temperature in Naturally ventilated buildings. It is calculated based on the Eq. (1) is used in the design and operation of naturally ventilated (NV) structures. It suggests that people in NV buildings thermally adjust to their surroundings' external temperature. It is calculated using the 30-day outdoor running mean temperature in degrees Celsius (NBC, 2016).

$$\text{Indoor operative temperature } T_n = (0.54 \times \text{outdoor temperature}) + 12.83 \quad (1)$$

where, outdoor temperature is obtained by taking the 30-day outdoor running mean air temperature (t_{rmt} in °C) and indoor operative temperature (T_n in °C) is the neutral temperature. Adaptive models specific for naturally ventilated structures in India have a 90% acceptance range of ± 2.38 °C (NBC, 2016).

The daily mean temperature (t_{m}) for a period of 30 days before the day of study. Running mean outdoor temperature of day 'n' (t_{rmt}) is calculated using the t_{m} values for the previous 30 days using the following Eq. (2) (Nicol et al., 2012):

$$t_{\mathrm{rmt}} = (1 - \alpha)\left\{t_{\mathrm{m},n-1} + \alpha t_{\mathrm{m},n-2} + \alpha^2 t_{\mathrm{m},n-3} + \ldots\right\} \quad (2)$$

The value of α is a number between 0 to 1 and a recommended value of 0.8 is taken for calculating t_{rmt} as prescribed in EN15251.

6 Quantitative Analysis

6.1 Building Selection

Two houses, one a vernacular house and the other a modern one are selected in the same village for the study. The vernacular house is representative one with all features of a vernacular construction techniques. The house is a 150-year-old structure built with thick kadapa stone wall masonry (Fig. 6). The house has an area of 59.55 m^2. The roof is made up of lime concrete laid over wooden rafters and flat kadapa stone slabs. It is a south-facing house with shared walls on the East and West. The house has a shaded veranda at the entrance and followed by an open hall and a kitchen.

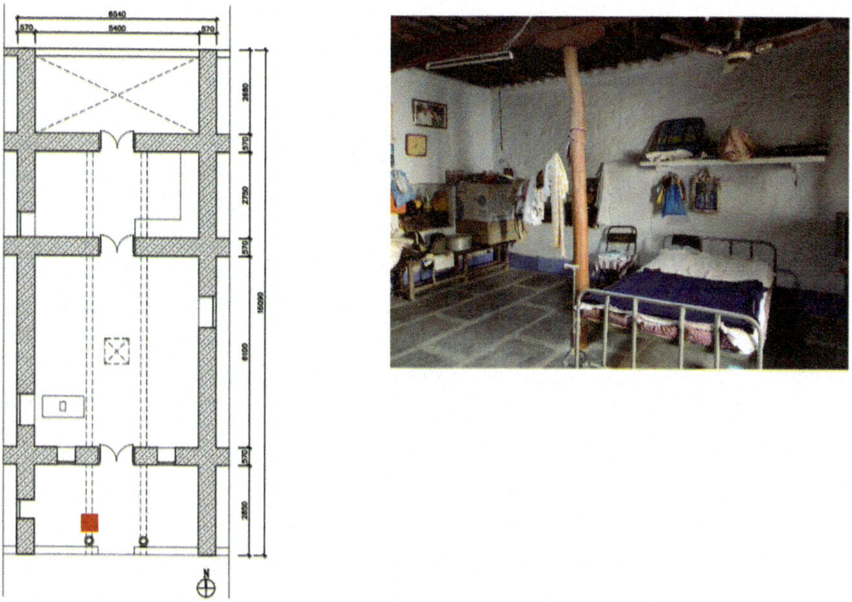

a) b)

■ Location of Data logger

Fig. 6 Vernacular house with kadapa-lime concrete roof; **a** floor plan **b** study area-hall interior

Kitchen further opens into an open courtyard on the North which houses the wet areas and toilets.

The house with RCC roof chosen for the study is a 2-bedroom house (Fig. 7). The walls are constructed with load-bearing kadapa stone masonry and cement mortar, whereas the roof is made of RCC slab. Much like the vernacular house with shared walls on the East and West, the room identified for the study has a shared wall on the west with the next house and on the east shares a wall with adjacent room. The house has an area of 59.27 m^2

6.2 *Experimental Setup and Field Measurements*

A temperature and humidity data logger TH32 of Tempnote make has been used to take readings of indoor temperature and relative humidity levels. The device is EN12830 certified and the resolution and accuracy levels are as per ASHRAE 55 standards (ASHRAE standard 55, 2020) (Table 1). The field measurements were taken on April 21st 2022 during summer. For a period of 24 h temperature and relative humidity within both dwellings are recorded at one-hour intervals. The device

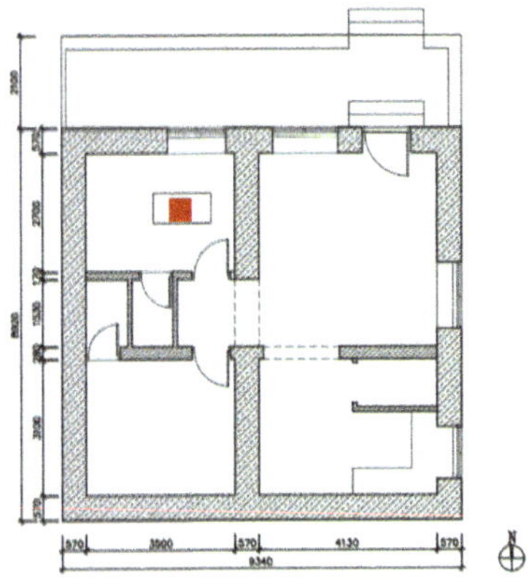
a)

b)

Location of Data logger

Fig. 7 Modern house with RCC slab; **a** floor plan **b** front view

Table 1 Survey instruments

Instruments and make	Climate parameter	Range	Resolution	Accuracy
Tempnote TH32 temperature	Air temperature	− 30 to + 60 °C	0.1 °C	± 0.5 °C
Humidity data logger	Relative humidity %	0% to 100% RH	0.1% RH	± 3% RH

is placed at a height of 1.1 m from ground level in both houses. The outdoor air temperature and relative humidity have been recorded.

6.3 *Questionnaire Survey*

On the same day as the field measurement, a questionnaire survey was conducted. The survey is designed to collect subjective responses of sensation votes from occupants on parameters like temperature, humidity, and overall thermal comfort. The selection of questions and components was based on a detailed review of literature (Dili et al., 2010; Nicol et al., 2012). The occupants are asked to assess the indoor environment without the use of any mechanical assistance to improve or modify thermal comfort. The questionnaire survey was prepared in English and Telugu, the native language. The questions are asked in the native language and the responses are recorded in a printed hand-out. 96 people were surveyed in total, 48 each from the vernacular houses and modern houses. Vernacular houses of about 150–180 years old are chosen for the survey. The modern houses selected were built in the last 20–30 years. The residents are informed about the scope and outcome of the research and is conducted with informed consent.

7 Results and Discussion

It is observed that the diurnal variation of the outdoor temperature is 13 °C, as it varied from a minimum of 28 °C to maximum of 41 °C. The indoor temperature in the vernacular house showed a diurnal variation of 2.6 °C, i.e., from 32.1 to 34.7 °C. The indoor temperature in the modern house showed a diurnal variation of 3.9 °C, i.e., from 33.4 to 37.3 °C. The peak outdoor temperature is 41 °C and the peak indoor temperature in the vernacular and modern house are 34.7 °C and 37.3 °C, respectively. The indoor temperatures of modern house are higher than the temperatures of vernacular house by 2.5–3 °C during the afternoon, whereas the temperature inside the vernacular house remained less than those of the modern house throughout the day (Fig. 8). The relative humidity inside the vernacular house varied from 41 to 64%, whereas in the modern houses it varied from 37 to 60% (Fig. 9). The lowest RH outside was 13% whereas, the lowest RH in the vernacular and modern houses are 41% and 39.3%, respectively.

Subjective responses to different thermal comfort parameters recorded in the questionnaire survey are illustrated below. The distribution graph of subjective responses on temperature in summer is shown in Fig. 10. Only 33% of the residents in the vernacular houses voted for warm and hot temperatures as against 100% in the modern houses for the same. 67% of the responses from the vernacular houses are spread between slightly cool to slightly warm. It is evident from the graph (Fig. 10)

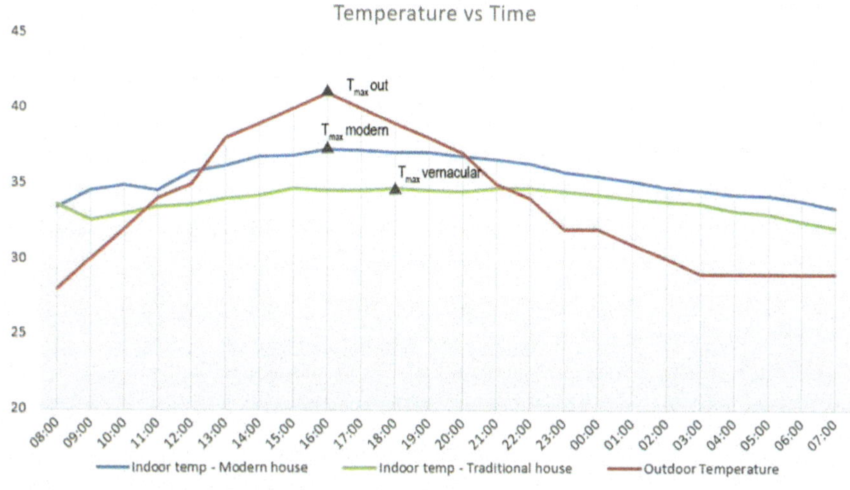

Fig. 8 Air temperature versus time

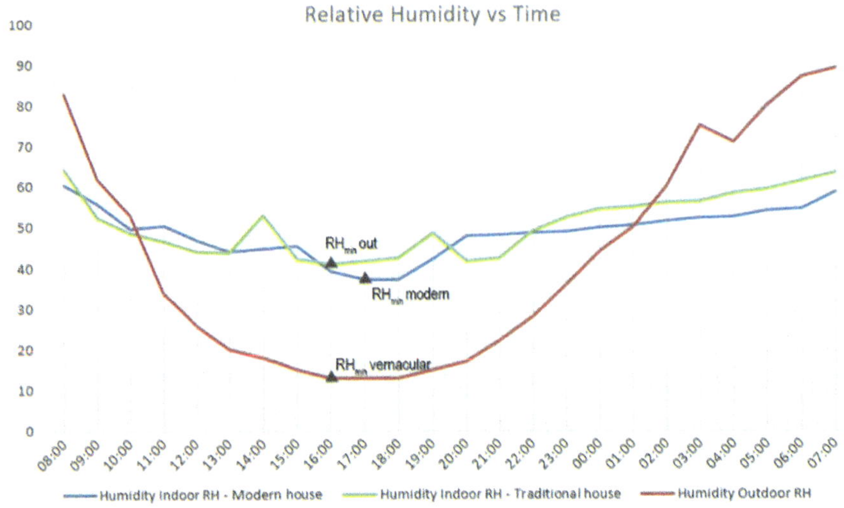

Fig. 9 Relative humidity % versus time

that no one from the modern houses feels their houses are slightly cool, neutral or slightly warm.

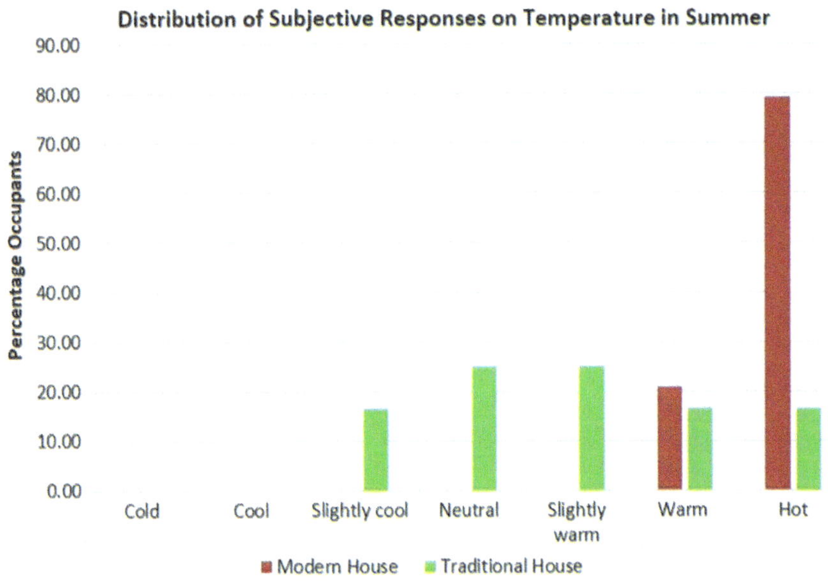

Fig. 10 Distribution of subjective responses on indoor temperature in summer

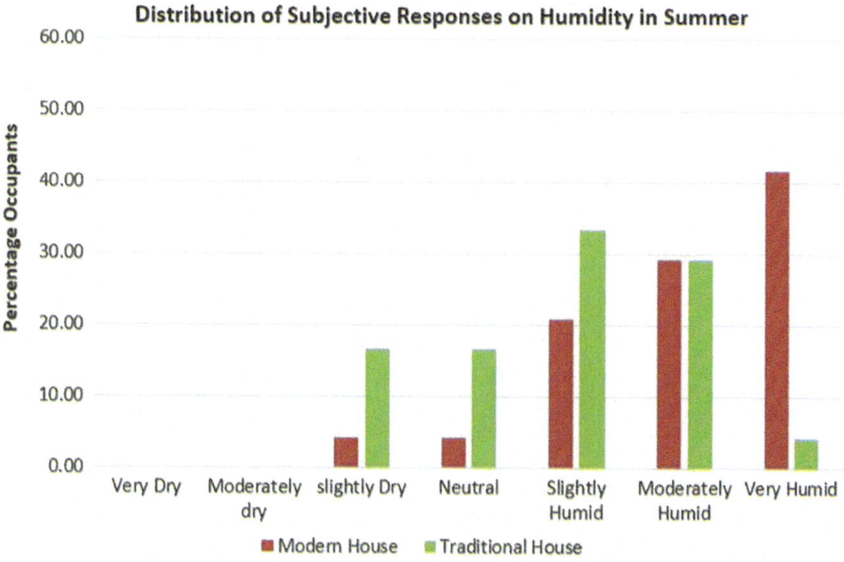

Fig. 11 Distribution of subjective responses on indoor humidity in summer

Figure 11 shows the distribution graph of subjective responses on indoor humidity in summer. 42% of the occupants from modern houses voted for very humid conditions while it is only 4% in vernacular houses. 63% and 50% from vernacular and modern houses, respectively, feel the conditions are slightly or moderately humid. 33% of the occupants from the vernacular houses feel the conditions are either slightly dry of neutral whereas only 8% feel the same in modern houses (Fig. 11).

Figure 12 shows the distribution graph of subjective responses on overall thermal comfort in summer. All the 48 occupants in modern houses voted for uncomfortable or very uncomfortable conditions and 43% of the occupants, which is 20 members from vernacular houses also feel the same. As much as 57% occupants in vernacular houses which is 28 members voted for slightly comfortable and comfortable conditions (Fig. 12). It is evident from the subjective responses that almost all of the occupants in modern houses did feel the conditions are uncomfortable, hot and humid, whereas the responses from vernacular houses are more distributed from slightly comfortable to uncomfortable conditions and are more towards the comfortable side.

8 Comparative Analysis Based on Adaptive Thermal Comfort Model

Comfort range for 90% acceptability for the summer month is calculated from the Eq. (1). The running mean outdoor temperature obtained from Eq. (2) is 34.25 °C and the calculated neutral temperature is 31.33 °C. For 90% acceptability, a limit of

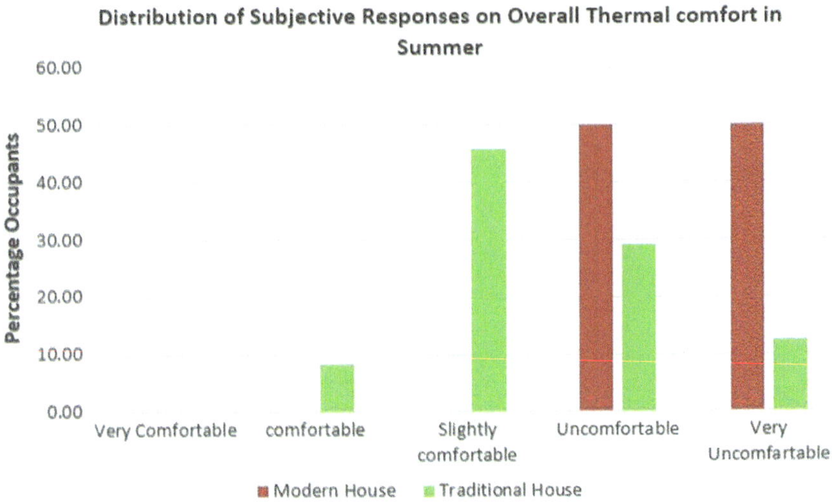

Fig. 12 Distribution of subjective responses on overall thermal comfort in summer

± 2.38 °C gives a comfort temperature range from 28.95 to 33.71 °C as shown on the graph in Fig. 13. The heating period, i.e., from sunrise to sunset is superimposed on the graph. Indoor relative humidity recorded for both the houses is also plotted on the graph. It is observed that the temperatures inside both the houses do not fall within the comfort range for most of the day. Temperatures inside the vernacular house fall with the comfort range from 2:00 a.m. to 12:00 p.m. Temperatures inside the modern house do not fall within the comfort range except for one hour in the morning between 6:00 a.m. to 7:00 a.m. However, it is observed that the temperatures inside the vernacular house are the closest to the comfort range with a deviation of only 0.1–1 °C above the upper comfort limit. Temperature inside the modern house has a deviation of 1–3.6 °C above the upper comfort limit (Fig. 13).

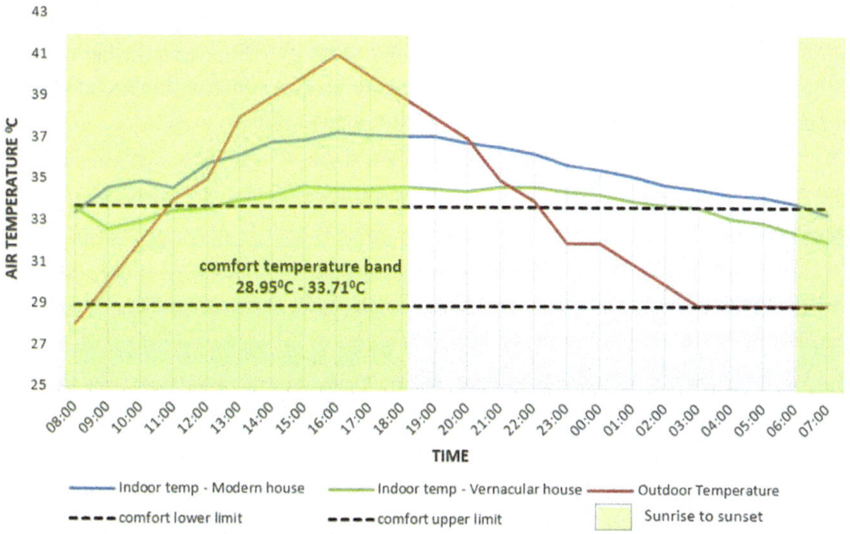

Fig. 13 Adaptive thermal comfort band and indoor air temperature

9 Conclusion

The indoor thermal comfort factors, methodologies, and construction practises used in vernacular homes in the Kadapa region for comfortable indoor conditions are investigated in this study. The major impact on indoor temperature is from roof. Temperatures inside the vernacular house are the closest to the comfort range with a deviation of only 0.1–1 °C above the upper comfort limit whereas the temperatures inside the modern house have a deviation of 1–3.6 °C above the upper comfort limit which is also evident from the responses from the questionnaire survey. It is understood that RCC may not be the right choice for roof construction in hot semi-arid conditions. The vernacular dwellings of Kadapa region with appropriate planning strategies and construction practices in response to local climate and material availability help in regulating the indoor air temperature by keeping it close to the comfort temperatures. The thick stone walls and roof have a high thermal capacity and can delay the heat gain indoors until late in the night as they can have higher time lag. The use of appropriate materials and traditional techniques in architecture can result in sustainable, energy-efficient, and comfortable indoor environments. The vernacular houses are load-bearing structures and may not be a feasible choice for multi-storey constructions, but the traditional techniques can be adopted to modern constructions for single residences to achieve comfortable indoor conditions. By documenting and validating the effectiveness of local, climate-responsive building techniques, the research offers actionable insights for reducing energy consumption, improving public health, and promoting sustainable infrastructure. It presents vernacular architecture as a viable solution to some of India's most pressing challenges, including energy efficiency, climate resilience, and environmental sustainability, thereby contributing to India's progress towards SDG 3, SDG 7, SDG 9, and SDG 13. This work further provides scope for developing a better alternative and vernacular construction techniques for a comfortable indoor environment in hot semi-arid regions.

References

ASHRAE Standard 55. (2020). Thermal comfort conditions for human occupancy. Atlanta.

Chellappa, V., & Srivastava, V. (2022). Science mapping approach based review of thermal comfort studies in Indian residential buildings. *Journal of Engineering, Design and Technology*.

Dili, A. S., Naseer, M. A., & Varghese, T, Z. (2010). Thermal comfort study of kerala vernacular residential buildings based on questionnaire survey among occupants of vernacular and modern buildings. *Energy and Buildings, 42*(11), 2139–2150.

ECBC. (2007). BEE.

Eco-Niwas Samhitha. (2018). BEE.

Joshima, V., Naseer, M., & Prabha. (2021). Assessing the real time thermal performance of reinforced cement concrete roof during summer—A study in the warm humid climate of Kerala. *Journal of Building Engineering*.

Kumar, S., Mathur, J., Mathur, S., & Singh, M. K. (2016). An adaptive approach to define thermal comfort zones on psychrometric chart for naturally ventilated buildings in composite climate of India. *Building and Environment, 109*, 135–153.

NBC. (2016). National building code.

Nicol, F., Michael, H., & Roaf, S. (2012). *Adaptive thermal comfort: Principles and practice*. CRC Press.

Vijayalaxmi, J., & Arathy, K. C. K. (2022). Critical evaluation of socio-cultural and climatic aspects in a traditional community: A case study of Pillayarpalayam weavers' cluster, Kanchipuram. *Built Heritage, 6*(1), 3.

Vijayalaxmi, J., & Sekar, S. P. (2013). Thermal performance of naturally ventilated residential buildings with various room orientations in the hot-humid climate of Chennai, India. *Journal of Architectural and Planning Research*, 1–22.

Assessment of Cross Laminated Timber with Coconut Wood as Primary Material

Aniket Naik and J. Vijayalaxmi

Abstract Alternative engineered wood products have great potential to replace concrete and steel in the structural application. The environmental footprint of cross laminated timber has been discussed widely as being advantageous. The aim of this study is to assess the reduction in embodied energy (EE) and Embodied Greenhouse Gas (EGHG) by using of Cross Laminated Coconut Timber (CLCT) as a structural component in a passive building as compared to the same building design using hypothetical conventional RCC structural system. The study will help in identifying the important strategies for the reduction of embodied energy and embodied greenhouse gas consumption at building level. Process analysis, which employs a bottom-up approach using physical quantities to estimate embodied energy is used to compare the embodied energy (EE) and embodied greenhouse gas (EGHG) consumption of Cross Laminated Coconut Timber (CLCT) and RCC Building. It is found that the cumulative embodied energy of the CLCT house is 291.11 GJ, while that of a similar reinforced cement concrete building is 399.06 GJ. The embodied energy of the CLCT building is 27% lesser than the RCC building. Further, the embodied greenhouse gas of the CLCT building is 65% lesser than a similar RCC building. Overall, in terms of embodied energy and embodied greenhouse gas consumption, CLCT building outperformed the RCC building.

Keywords Coconut wood · Cross laminated timber · Embodied energy · Embodied greenhouse gas

Nomenclature

CLT Cross Laminated Timber

A. Naik (✉)
Environmental Design Solutions Pvt. Ltd., New Delhi, India
e-mail: aniketnaik297@gmail.com

J. Vijayalaxmi
School of Planning and Architecture, Vijayawada, India
e-mail: vijayalaxmij@spav.ac.in

J. Vijayalaxmi (ed.), *Sustainable Building Performance*,
https://doi.org/10.1007/978-981-96-2428-7_4

CLCT	Cross Laminated Coconut Timber
EE	Embodied Energy
EGHG	Embodied Greenhouse Gas
VOC	Volatile Organic Compound
LCA	Life Cycle Assessment
LCGHGE	Life Cycle Greenhouse Gas Emissions

1 Introduction

Building professionals worldwide are increasingly considering alternatives to timber because of a relatively small manufacturing energy input and the perception that when substituted for other materials, timber has carbon benefits. For sustainable construction of residential and mixed-use buildings, researchers have identified Cross laminated timber (CLT) as an emerging structural system. CLT construction has already been introduced in many developed countries and there is a need to introduce it in India at a larger scale. Although no tall CLT structure has been erected in India, local producers of CLT panels have established themselves, and the CLT components are being used as floor, roof, or structural components in several projects. Since the 1990s, a number of research on CLT shear walls, building assembly, and environmental impact have been undertaken in developed countries. However, no studies have evaluated the environmental impact of the development of CLT buildings in India (Akiner et al., 2022; Balguri et al., 2021; Bhardwaj et al., 2023; Budarapu et al., 2016; Kumar & Allamraju, 2019).

Recently, there have been attempts to use CLT products from coconut trunk in mass-timber building construction in Goa. However, there is limited research available in India on the structural, thermal performance, cost analysis, EE, and GHG emissions of CLCT construction systems. This emphasizes the need to undertake a study to assess embodied energy and greenhouse gas assessment of CLT structures in order to influence the selection of environment-friendly materials and construction techniques (Budarapu et al., 2014; Jaffery et al., 2019; Jha et al., 2022; Liisma et al., 2019; Murali Mohan et al., 2021; Sudhirsastry et al., 2015).

Construction processes and the usage of building products account for a significant share of human-caused global environmental impacts. Buildings account for 36% of global energy usage, 40% of global material use, and 39% of global energy-related greenhouse gas emissions, International Energy Agency and United Nations Environment Program, all of which have severe environmental consequences. As a result, there is an urgent need to adapt sustainable construction techniques through the use of high-performance materials with low-carbon footprint. CLT is composed of three, five, or seven layers of timber planks that are bonded together at high pressures and oriented in orthogonal directions. The LCA and LCGHGE of CLT and RCC buildings are influenced by various factors, including raw material production, building service life and operations, and end-of-life strategies (Dinbandhu et al., 2021a, 2021b;

Jayalath et al., 2020; Korpi et al., 2019; Santamouris & Vasilakopoulou, 2021; Yadav et al., 2014).

Pei provided a detailed overview of the engineering and social benefits and challenges of CLT construction in the United States. According to this analysis, the most economically viable market for CLT is between 8 and 15 floors, where standard light-framed wood systems are not feasible owing to the risk of fire. Although a number of studies on the structural performance of CLT buildings have been done in the United States and Canada, the environmental performance in terms of its EE and EGHG of this innovative building system has not been thoroughly examined, especially in India (Gupta et al., 2019; Dinbandhu et al., 2021b; Krishnaja et al., 2018; Pei et al., 2013, 2016).

Schneider et al., (2014) conducted a building energy consumption analysis on a CLT building (a four-story multi-residential building) and published the results in the CLT Handbook The results show a 14% potential energy cost savings when compared to a light timber framed building (which is less thermally massive than a CLT building) within the Sacramento, California, climate. This study can be used as base evidence that tall CLT construction can save energy, but further research is needed to gain trust in this innovative building method (Gagnon et al., 2013; Popovski et al., 2012; Schneider et al., 2014). Robertson et al. (2012) used primary data from a life cycle inventory at a CLT factory in British Columbia to perform a comparative LCA analysis on a five-story RCC framed office structure and a CLT and Glulam hybrid building. In nine out of eleven environmental indices, the CLT and Glulam hybrid building has a lower environmental effect than the concrete building.

Several studies on the configuration and mechanical properties (Mayencourt & Mueller, 2019; Wang & Yin, 2022), production, characteristic material properties, design (Brandner et al., 2016), economic viability of CLCT (Smyth, 2018), design methods and concepts (Schenk et al., 2022), energy reductions in Nordic countries and Europe (Svortevik et al., 2020; Younis & Dodoo, 2022) have been carried out. However, studies on the environmental repercussions of CLCT as a structural component in India are limited. Hence this study aims to assess the embodied energy and Embodied Green House Gas impacts of using CLCT as compared to a hypothetical reinforced cement concrete building of similar design. This study gains further significance considering the extensive availability of Coconut Timber in South India. Some other researchers work on different materials like isotopes of precipitation have climate change, pedogenic processes in loess-paleosol sediments, modeling of multi row stabilizing and carbon fiber reinforced plastic plate (Monteiro et al., 2016; Zhang et al., 2023a, 2023b; Zhu et al., 2023).

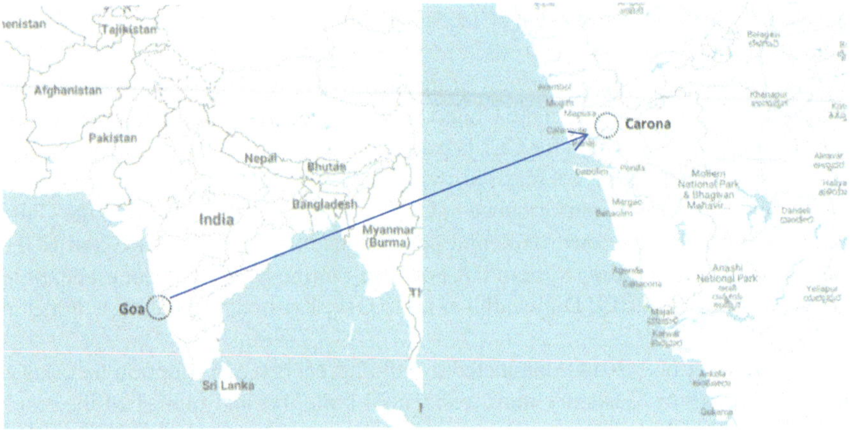

Fig. 1 Case study location (*Source* Author based on Snazzymaps 2022)

2 Materials and Methods

2.1 Study Area

This case study investigates the EE and EGHG of a low-rise CLT building in Carona Village, Goa, India (Snazzymaps, 2022) with a built-up area of 246.6 m^2 as seen in Fig. 1. The evaluation of different design scenarios is centered on two important systems: primary structural support (CLT or RCC) and building enclosure system.

2.2 Case Study Building Description

The case study building (Fig. 2) is a permanent structure made using salvaged coconut wood from trees which are cut after surpassing their productive life (Charuau, 2021). The butterfly roof sits on the timber frame, anchored to the ground by steel joints that transfer the weight of the superstructure to the masonry plinth in the rear portion of the house and concrete column in the front which can be clearly noted in Fig. 3 (Charuau, 2021). The structural frame is designed on a horizontal grid of 1.9 m to avoid wastage of the infill material and the plywood (Fig. 4). Coconut battens create a grillwork that allows light and breeze throughout the house.

CLCT columns and beams of 160 mm width and 65 mm thickness are used for the structural system. The roof is supported on seven sets of sandwiched double CLCT beams of 65 mm thickness. The CLCT building's roof is made of 25 mm thick CLT boards with corrugated metal covering. There are 79 CLT Columns and 279 CLT Beams (103 Primary Beams and 169 Secondary Beams) as shown in Fig. 4. The structural members are shown in the 3D model generated using Sketchup Trimble.

Fig. 2 Case study house taken for study (*Source* Charuau 2021)

Fig. 3 Roof system using CLT in the case study (*Source* Charuau 2021)

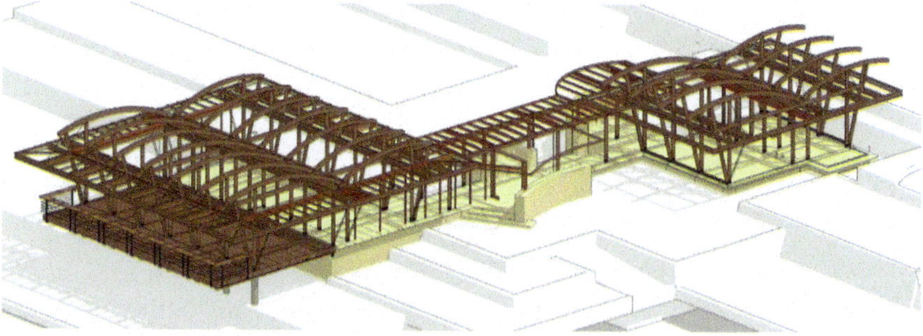

Fig. 4 3D model of the structural system using CLT developed in SketchUp trimble (*Source* author)

For the same building plan, when RCC structures are used as done conventionally, the columns are 200 mm thick and beams of 300 mm width and 450 mm depth are used. A total of 57 RCC Columns and 154 RCC Beams (57 Primary and 97 Secondary Beams) are arrived as per the structural load. Plinth beams of 200 mm × 400 mm are used. The roof structure of a conventional RCC building would be made up of a 150 mm thick concrete slab.

It should be noted that in the comparison of embodied energy and embodied GHG emissions, the building components that are common to the field case study of CLT and the hypothetical RCC building for the same plan design such as doors, windows, and lovers are not considered. The plan of the field study CLT building and the conventional building using RCC components is the same and shown in Fig. 5. The pictures of the case study room are shown in Fig. 6 and the sections are shown in Figs. 7 and 8.

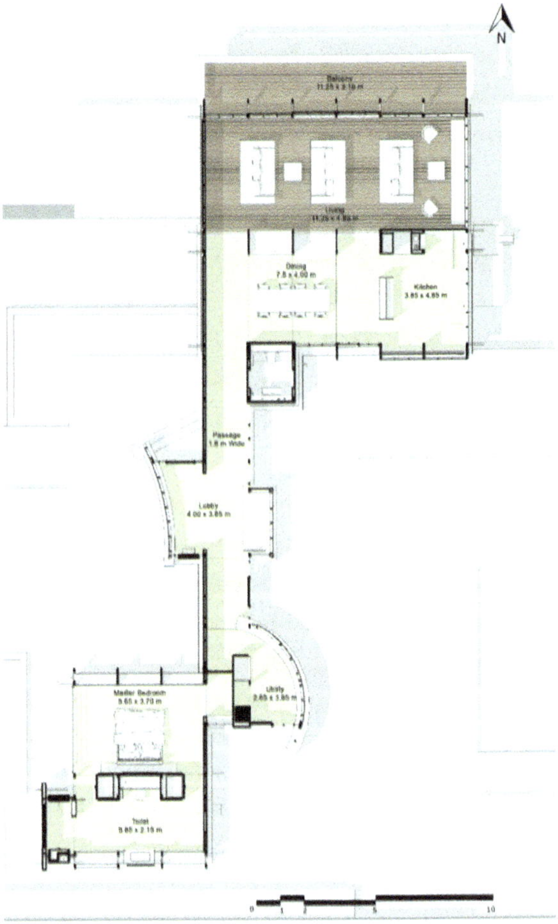

Fig. 5 Floor plan of the case study building—CLT structure and RCC structure (*Source* author)

Fig. 6 Bedroom space in the field case study CLT building (*Source* Charuau 2021)

Fig. 7 Section through the living room and passage of the CLT field case study (*Source* author)

Fig. 8 Section through the bedroom of the case study (*Source* author)

2.3 Methodology

The case study building is built using CLT. It is also redesigned using the conventional structural system of reinforced cement concrete system for the same plan in order to conduct a comparative analysis of the two alternative construction scenarios. The goal is to develop a functionally similar structural system and building components rather than execute a direct material substitution or replacement. Using the original plans and elevations of the CLT building an equivalent representative reinforced cement concrete building is designed.

2.3.1 Embodied Energy

The construction sector is a large user of primary energy resources, accounting for up to 20–25% of India's overall energy consumption (Monahan & Powell, 2011). With the world's energy situation unclear, buildings should be planned and built to consume less fossil fuel-based energy and have reduced environmental implications (Azari & Abbasabadi, 2018). A significant quantity of energy is spent on the embodied energy of the buildings. Some studies comprehensively define that embodied energy comprises the energy consumed during the extraction and processing of raw materials, transportation of the original raw materials, manufacturing of building materials and components and energy use for various processes during the construction and demolition of the building (Cabeza et al., 2013; Ding, 2005). The embodied energy calculation can be done for a single building to gauge the efficiency of the dominant building material used (Chel & Tiwari, 2009; Shukla et al., 2009) or for multiple houses for a comparative study.

More recent studies have been focused on comparative analysis of materials. With the introduction of alternative materials, some studies comparing conventional and alternative building materials have been conducted to point out the advantages and justify the usage of alternative building materials (Cherian et al., 2020; Dabaieh et al., 2020). One of the studies focused on the difference in embodied energy for a new and an existing building (Monteiro et al., 2016).

2.3.2 Embodied Green House Gas

Embodied GHG emissions occur due to the burning of fossil fuels during extraction, manufacture, electricity consumption for other purposes, transportation, and erection of a material in total (Simonen et al., 2017). Buildings consume around 32% of global final energy. They were also responsible for the emission of 32% of global GHG (IPCC, 2014). Nearly 50% of these emissions are attributed to building materials such as iron, cement, and steel. The Embodied GHG emissions due to its building materials can account for up to 75% over the building's lifetime (Kristjansdottir et al., 2018).

India's CO_2 emissions have become thrice as compared to its emissions in 1990. India seeks to reduce its CO_2 emissions by at least 35% to meet COP26 commitment of Paris Agreement (Singh Ahluwalia & Patel, 2021).

A study on the GHG emissions of office buildings in the United States showed that steel buildings of five floors emitted 54% more GHG than wood buildings (Gursel et al., 2023). This study also estimated that the U.S. had a total CO_2 eq emission of nearly 138 million metric tons due to the use of steel, concrete, and wood. Such studies are important as they give the designers an opportunity to make early design stage environmentally favorable decisions.

2.3.3 Study Method

In this study, the EPiC database is used to obtain the Embodied Energy and Embodied GHG emission coefficients for the materials used (Crawford et al., 2020).

A bill of materials is computed for both CLCT and RCC design alternatives for the structural component. The details required to calculate the material quantities of the existing case study are obtained from the architectural drawings of the CLCT building, bill of quantities, and technical specifications provided by the architect. This bill of construction materials is used to assess embodied energy and greenhouse gas emissions. For comparative analysis, total embodied energy and embodied greenhouse gas estimations are performed and classified for each possible design scenario.

The embodied energy coefficients may be calculated using a number of methodologies, and the results can vary (Cabeza et al., 2014). The most frequently used method for LCEA is Process analysis, which employs a bottom-up approach using physical quantities to estimate embodied energy.

To conduct a comparative analysis, the total embodied energy and embodied greenhouse gas emissions are estimated and classified for the existing CLCT building and the RCC design case. The data obtained from the EPiC database, combined with the bill of materials, allows for a comprehensive evaluation of the environmental impact associated with each construction scenario.

Given the availability of context-specific data, this study is designed to evaluate the embodied energy (EE) and Embodied greenhouse gas emissions (EGHG) of CLCT and RCC construction specifically in India.

3 Observation and Analysis

3.1 Embodied Energy and Embodied GHG Emissions

The total energy spent during the extraction of raw materials, processing, production, and transportation is accounted for by the initial embodied energy. The top-down approach of the input–output technique is based on energy intensity by fuel type and economic data. Use of national average pricing, translation of economic data into energy data, homogeneity assumptions, and data age are all constraints of the input–output approach (Crawford, 2008).

Hybrid techniques combine these two sources, combining accessible process data with input–output data across the goods and services supply chain (Chastas et al., 2017). Some other researchers work on different materials like Zhu et al. (2023) isotopes of precipitation have climate change, He et al. (2021), pedogenic processes in loess-paleosol sediments, Zhang et al. (2022) modeling of multi-row stabilizing and Zhang et al. (2023b) carbon fiber reinforced plastic plate.

Table 1 EE and EGHG intensities of materials

Material	Density (kg/m^3)	Embodied energy (MJ/m^3)	Unit	Embodied GHG (kg CO_2 eq/m^3)	Unit
Cross laminated timber (CLT)	490	9607	MJ/m^3	645	kg CO_2 eq/m^3
Reinforced concrete (RC)	2409	2581	MJ/m^3	361	kg CO_2 eq/m^3

Source EPiC database (Gursel et al., 2023)

In comparison to process analysis, the hybrid method yields greater estimates of embodied energy. Equation 1 is used to determine the initial embodied energy of the two case study buildings, and hybrid energy coefficients from the EPiC database are used to calculate the embodied energy (Crawford et al., 2020).

Equation 1

$$\mathrm{EE} = \sum_{m=1}^{M} Q_m \mathrm{EE}_m \tag{1}$$

where EE (GJ) denotes total embodied energy, Qm (kg, m^3, m^2) denotes material m quantity, and EE (GJ/unit) denotes material m's embodied energy coefficient. This study does not take into account operational embodied energy related to building retrofit and maintenance.

Equation 2 is used to calculate the emissions associated with the manufacture of construction materials using the embodied GHG intensities indicated in Table 1.

Equation 2

$$\mathrm{GHG}_M = \sum_{m=1}^{M} Q_m \mathrm{GHG}_m \tag{2}$$

where GHG_M (kg CO_2-e) denotes the total embodied GHG of the building and GHG_m (kg CO_2-e per unit) denotes the embodied GHG coefficient of material *m*.

Table 1 specifies the Density, EE intensity, and EGHG intensities whereas Tables 2 and 3 depict the breakdown of EE during the CLCT building's and RCC building's construction stage, respectively.

4 Results and Discussions

Figure 9 depicts the breakup of embodied energy in the CLCT and RCC structures during construction. The CLCT building is used as the baseline, and its total embodied energy is found to be 37% lower than that of the RCC structure as shown in Table 4.

Table 2 EE intensity and EGHG intensity breakdown in Case 1—CLCT building

Components	Volume (m^3)	Density (kg/m^3)	Mass (kg)	Embodied energy (MJ/Unit)	Unit	Total embodied energy (MJ)	Embodied GHG (kg CO_2 eq/unit)	Unit	GHG emissions (kg CO_2 eq)
Structure									
Columns	2.92	490	1431.01	9607	m^3	28,056.51	645	m^3	1883.67
Beams	8.54	490	4184.26	9607	m^3	82,037.17	645	m^3	5507.86
Floor									
Floor Beams	4.05	490	1984.40	9607	m^3	38,906.42	645	m^3	2612.12
Floor Planks	3.53	490	1730.21	9607	m^3	33,922.70	645	m^3	2277.52
Roof									
Ceiling Planks	11.26	490	5518.06	9607	m^3	108,187.67	645	m^3	7263.56
			Total	Total		291,110.48			19,544.73

Table 3 EE intensity and EGHG intensity breakdown in Case 1—RCC building

Components	Volume (m^3)	Density (kg/ m^3)	Mass (kg)	Embodied energy (MJ/ Unit)	Unit	Total embodied energy (MJ)	Embodied GHG (kg CO_2 eq/unit)	Unit	GHG Emissions (kg CO_2 eq)
Structure									
Columns	13.22	2409	31,858.06	2581	m^3	34,132.69	361	m^3	4774.08
Beams	50.04	2409	120,549.06	2581	m^3	129,156.13	361	m^3	18,064.84
Floor									
Plinth Beams	10.54	2409	25,396.64	2581	m^3	27,209.93	361	m^3	3805.81
PCC	13.24	2409	31,898.53	2581	m^3	34,176.05	361	m^3	4780.15
Roof									
Slab	67.57	2409	162,771.38	2581	m^3	174,393.08	361	m^3	24,392.06
			Total	Total		399,067.89			55,816.93

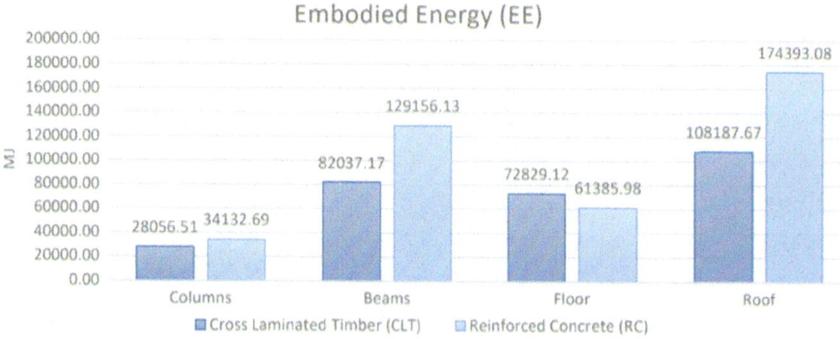

Fig. 9 Comparison of embodied energy of components of CLT and RCC building

The breakup of components in Fig. 9 shows that the variation between the RCC and CLCT building is due to the main structural elements in the building, which include columns, beams, floor, and roof.

This is mostly due to the high mass of energy-intensive materials used in RCC construction. The use of CLCT as a structural component results in considerable embodied energy reductions. It should be emphasized that while the floor component of a CLCT structure has a higher embodied energy than an RCC building, it only accounts for a small fraction of the total embodied energy. The roof component of the RCC building shows the highest embodied energy followed by the beam component.

Figure 10 shows the breakup of embodied GHG emissions from the CLCT and RCC building construction phases. It can be seen that higher energy intensive processes, as well as a large volume of materials and GHG emissions, result in a significant increase in the RCC building's total Embodied GHG emissions. Roof component of the RCC building has the highest embodied energy greenhouse gas emissions, followed by the beam component. The structural elements of both structures account for the majority of the material volumes utilized. The increase in Embodied GHG emissions of the RCC building is 186 percent due to high GHG emission coefficients with high density of concrete compared to CLCT.

When analyzing the embodied energy component wise, the EE of Columns in the RCC building is found to be 21.6% higher than that of the CLT building, while beams in RCC building consume 57.4% more of EE. Surprisingly, the EE of flooring element in RCC building is lower than the CLT by 15%. This is because, the integrated structure of columns and plinth beams in RCC building facilitates use of lesser floor

Table 4 Comparison of total EE and EGHG in Case 1 and Case 2

Energy	Case 1—CLT building	Case 2—RC building	Variation (%)
Embodied energy (EE) MJ	291,110.48	399,067.89	27
Embodied greenhouse gas (EGHG) $kgCO_2eq$	19,544.73	55,816.93	65

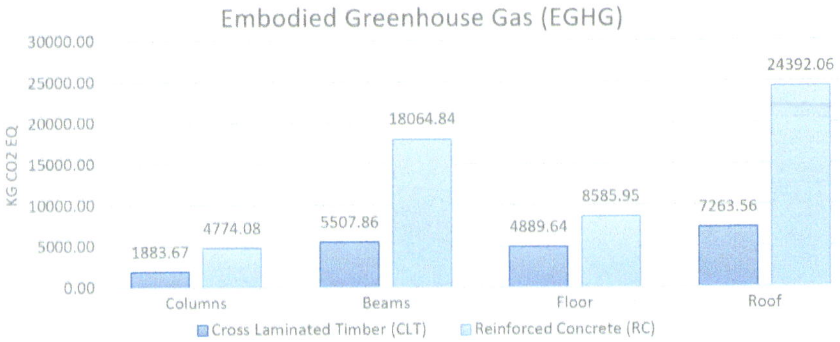

Fig. 10 EGHG comparison of both the cases—CLCT and RCC building

beams, which is not so in the case of the CLT building. The embodied energy of the ceiling is 61% higher than the CLT building.

This shows that there is ample opportunity to use CLT in Roofing of RCC building in order to reduce its overall EE.

The results show huge increase in Embodied Green House Gas for all the components of the RCC building. Columns in RCC building emit 153% of EGHG as compared to the CLT building, while beams emit 228% as compared to CLT buildings. Floor beams of RCC emit 75.5% more GHG while ceiling emits 236% of GHG.

From Fig. 11 it can be seen that the percentage of GHG in the CLT and RCC building is similar. Use of CLT greatly reduces the GHG emissions in every component of the building. For all practical purposes, CLT can be used in part of the RCC buildings wherever possible to reduce the GHG.

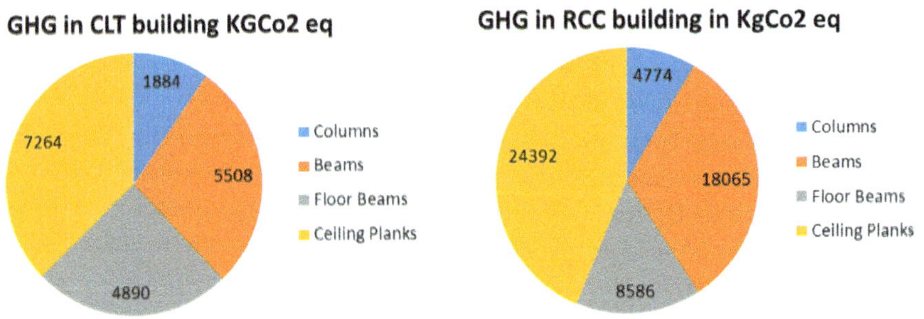

Fig. 11 GHG emission of four major structural components in CLT building and RCC building

5 Conclusion

This study analyses the embodied energy and embodied greenhouse gas emissions for two types of low-rise residential structures made with CLT and RCC. The real-life case study of a building constructed using CLT is analyzed and the same building is redesigned using RCC, as it is the conventional material in this part of the world. The RCC building's embodied energy and embodied GHG emissions are 27 percent and 65 percent greater, respectively, than the CLCT building. The embodied Energy of all structural components in the CLT building is lower than the RCC building, except the flooring. This indicates a need to further research on alternative structural systems in CLT buildings for flooring component. The Embodied Green House Gas emission of the RCC building is high for all the structural components. Columns, Beams, Floor beams, Ceiling in RCC building emit 153%, 228%, 75.5%, and 236% of GHG as compared to the CLT building. This shows the low energy intensity of CLT and it can be used as an eco-friendly material, India has a long coast and coconut trees are available in plenty. Hence CLT can be used with coconut timber as a primary material.

The net benefits of CLT construction outweigh those of RCC construction, which is in line with previous research findings. In the construction stage of RCC building, high quantity of energy-intensive materials with higher embodied energy and embodied GHG coefficients derived from the hybrid technique has increased embodied energy and embodied GHG emissions. The CLT building in this study has reduced embodied energy and embodied greenhouse gas emissions, indicating that the structural system and enclosure components of the building play a significant role in lowering embodied energy and greenhouse gas emissions. CLT shows potential as a structural material with reduced embodied energy and embodied greenhouse gas emissions while also being a lightweight, carbon-storing, and renewable alternative to concrete. However, the CLT building requires more maintenance, termite proofing, and other high-volume building components (e.g., plasterboard), all of which can result in greater recurring and operational GHG emissions. Moreover, the performance of CLT structures must be approved in India by regulatory agencies and market and be used in the mainstream practice. Factors affecting livability, comfort, and the resulting socio-economic consequences will have an impact on societal acceptance of CLT buildings from the user's perspective. Further research could explore the effects of operational embodied energy considering the building's service life.

This study is essential for making informed decisions regarding construction materials and practices, as it provides insights into the energy intensity associated with different construction techniques. Understanding the significance of embodied energy allows researchers, architects, and builders to identify materials with lower embodied energy, promoting the adoption of more sustainable and environmentally friendly alternatives.

Acknowledgements The authors thank Ini Chatterji & Associates for having shared the drawings which enabled Figs. 4, 5, 7, and 8.

Author Contributions All authors have equally contributed to the research work.

Conflict of Interest The authors declare no conflict of interest.

Data Availability All the data are available in the manuscript.

References

Akiner, M. E., Akiner, İ, Akiner, N., & Pancovska, V. Z. (2022). Using wood as a new generation building material in the context of sustainable development. *Zaštita Materijala, 63*, 68–78. https://doi.org/10.5937/ZASMAT2201068A

Azari, R., & Abbasabadi, N. (2018). Embodied energy of buildings: A review of data, methods, challenges, and research trends. *Energy and Buildings, 168*, 225–235. https://doi.org/10.1016/J.ENBUILD.2018.03.003

Balguri, P. K., Samuel, D. G. H., & Thumu, U. (2021). A review on mechanical properties of epoxy nanocomposites. *Materials Today: Proceedings, 44*, 346–355. https://doi.org/10.1016/J.MATPR.2020.09.742

Bhardwaj, A. R., Vaidya, A. M., Meshram, P. D., & Bandhu, D. (2023). Machining behavior investigation of aluminium metal matrix composite reinforced with TiC particulates. *International Journal on Interactive Design and Manufacturing, 2023*, 1–15. https://doi.org/10.1007/S12008-023-01378-6

Brandner, R., Flatscher, G., Ringhofer, A., Schickhofer, G., & Thiel, A. (2016). Cross laminated timber (CLT): Overview and development. *European Journal of Wood and Wood Products, 74*, 331–351. https://doi.org/10.1007/S00107-015-0999-5/METRICS

Budarapu, P. R., Sudhir Sastry, Y. B., Javvaji, B., & Mahapatra, D. R. (2014). Vibration analysis of multi-walled carbon nanotubes embedded in elastic medium. *Frontiers of Structural and Civil Engineering, 8*, 151–159. https://doi.org/10.1007/S11709-014-0247-9/METRICS

Budarapu, P. R., Sudhir Sastry, Y. B., & Natarajan, R. (2016). Design concepts of an aircraft wing: Composite and morphing airfoil with auxetic structures. *Frontiers of Structural and Civil Engineering, 10*, 394–408. https://doi.org/10.1007/S11709-016-0352-Z/METRICS

Cabeza, L. F., Barreneche, C., Miró, L., Morera, J. M., Bartolí, E., & Inés Fernández, A. (2013). Low carbon and low embodied energy materials in buildings: A review. *Renewable and Sustainable Energy Reviews, 23*, 536–542. https://doi.org/10.1016/J.RSER.2013.03.017

Cabeza, L. F., Rincón, L., Vilariño, V., Pérez, G., & Castell, A. (2014). Life cycle assessment (LCA) and life cycle energy analysis (LCEA) of buildings and the building sector: A review. *Renewable and Sustainable Energy Reviews, 29*, 394–416. https://doi.org/10.1016/J.RSER.2013.08.037

Charuau, F. (2021). [Pictures of Ini Chatterjee home]. [Photograph] STUDIO Charuau. https://www.fabiencharuauphotography.com/ini-chatterjee-home

Chastas, P., Theodosiou, T., Bikas, D., & Kontoleon, K. (2017). Embodied energy and nearly zero energy buildings: A review in residential buildings. *Procedia Environmental Sciences, 38*, 554–561. https://doi.org/10.1016/J.PROENV.2017.03.123

Chel, A., & Tiwari, G. N. (2009). Thermal performance and embodied energy analysis of a passive house—Case study of vault roof mud-house in India. *Applied Energy, 86*, 1956–1969. https://doi.org/10.1016/J.APENERGY.2008.12.033

Cherian, P., Palaniappan, S., Menon, D., & Anumolu, M. P. (2020). Comparative study of embodied energy of affordable houses made using GFRG and conventional building technologies in India. *Energy Buildings, 223*, 110138. https://doi.org/10.1016/J.ENBUILD.2020.110138

Crawford, R. H. (2008). Validation of a hybrid life-cycle inventory analysis method. *Journal of Environmental Management, 88*, 496–506. https://doi.org/10.1016/J.JENVMAN.2007.03.024

Crawford, R. H., Stephan, A., & Prideaux, F.: EPiC: Introducing a database of hybrid environmental flow coefficients for construction materials. *IOP Conference Series Earth and Environmental Science, 588*. https://doi.org/10.1088/1755-1315/588/4/042052

Dabaieh, M., Heinonen, J., El-Mahdy, D., & Hassan, D. M. (2020). A comparative study of life cycle carbon emissions and embodied energy between sun-dried bricks and fired clay bricks. *Journal of Cleaner Production, 275*, 122998. https://doi.org/10.1016/J.JCLEPRO.2020.122998

Dinbandhu, Abhishek, K., Thakur, A., Nagaphani Sastry, M., Devaki Devi, K., & Nishant, A. (2021a). A study on mechanical attributes of epoxy-carbon fiber-terminalia bellirica embedded hybrid composites. In *Lecture notes in mechanical engineering* (pp. 163–173). Springer. https://doi.org/10.1007/978-981-16-3033-0_15

Dinbandhu, Thakur, A., Venugopal Goud, E., Abhishek, K., & Vora, J. J. (2021b). An overview of Proteus: The world's first man-made non-cuttable material. In *Lecture notes in mechanical engineering* (pp. 95–102). Springer. https://doi.org/10.1007/978-981-16-3033-0_9

Ding, G. K. C. (2005). Developing a multicriteria approach for the measurement of sustainable performance. *Building Research and Information, 33*, 3–16. https://doi.org/10.1080/096132104 2000322618

Gagnon, S., Bilek, E., et al. (2013). CLT: Introduction to cross-laminated timber. fpl.fs.usda.gov.

Gupta, T. K., Budarapu, P. R., Chappidi, S. R., YB, S. S., Paggi, M., & Bordas, S. P. (2019). Advances in carbon based nanomaterials for bio-medical applications. *Current Medicinal Chemistry, 26*, 6851–6877. https://doi.org/10.2174/0929867326666181126113605

Gursel, A. P., Shehabi, A., & Horvath, A. (2023) Embodied energy and greenhouse gas emission trends from major construction materials of U.S. office buildings constructed after the mid-1940s. *Building and Environment, 234*, 110196. https://doi.org/10.1016/J.BUILDENV.2023.110196.

He, M., Dong, J., Jin, Z., Liu, C., Xiao, J., Zhang, F., Sun, H., Zhao, Z., Gou, L., Liu, W., Luo, C., Song, Y., Ma, L., & Deng, L. (2021). Pedogenic processes in loess-paleosol sediments: Clues from Li isotopes of leachate in Luochuan loess. *Geochimica et Cosmochimica Acta, 299*, 151–162. https://doi.org/10.1016/j.gca.2021.02.021

IPCC: Climate Change 2014: Mitigation of climate change. Summary for policymakers and technical summary. Climate Change 2014: Mitigation of Climate Change: Working Group III Contribution to the IPCC Fifth Assessment Report (pp. 1–161). https://doi.org/10.1017/CBO978110 7415416.005

Jaffery, H. A., Sabri, M. F. M., Said, S. M., Hasan, S. W., Sajid, I. H., Nordin, N. I. M., Megat Hasnan, M. M. I., Shnawah, D. A., & Moorthy, C. V. (2019). Electrochemical corrosion behavior of Sn-0.7Cu solder alloy with the addition of bismuth and iron. *Journal of Alloys and Compound, 810*, 151925. https://doi.org/10.1016/J.JALLCOM.2019.151925

Jayalath, A., Navaratnam, S., Ngo, T., Mendis, P., Hewson, N., & Aye, L. (2020). Life cycle performance of cross laminated timber mid-rise residential buildings in Australia. *Energy and Buildings, 223*, 110091. https://doi.org/10.1016/J.ENBUILD.2020.110091

Jha, P., Shaikshavali, G., Shankar, M. G., Ram, M. D. S., Bandhu, D., Saxena, K. K., Buddhi, D., & Agrawal, M. K. (2022). A hybrid ensemble learning model for evaluating the surface roughness of AZ91 alloy during the end milling operation. *Surface Review and Letters*. https://doi.org/10.1142/S0218625X23400012

Korpi, A. G., Ţălu, Ş, Bramowicz, M., Arman, A., Kulesza, S., Pszczolkowski, B., Jurečka, S., Mardani, M., Luna, C., Balashabadi, P., Rezaee, S., & Gopikishan, S. (2019). Minkowski functional characterization and fractal analysis of surfaces of titanium nitride films. *Materials Research Express, 6*, 086463. https://doi.org/10.1088/2053-1591/AB26BE

Krishnaja, D., Cheepu, M., & Venkateswarlu, D. (2018). *A review of research progress on dissimilar laser weld-brazing of automotive applications*. https://doi.org/10.1088/1757-899X/330/1/012073

Kristjansdottir, T. F., Heeren, N., Andresen, I., & Brattebø, H. (2018). Comparative emission analysis of low-energy and zero-emission buildings. *Build Research and Information, 46*, 367–382. https://doi.org/10.1080/09613218.2017.1305690

Kumar, P. S. S., & Allamraju, K. V. (2019). A review of natural fiber composites [Jute, Sisal, Kenaf]. *Materials Today: Proceedings, 18*, 2556–2562. https://doi.org/10.1016/J.MATPR.2019.07.113

Liisma, E., Kuus, B. L., Kukk, V., & Kalamees, T. (2019). A case study on the construction of a CLT building without a preliminary roof. *Journal of Sustainable Architecture and Civil Engineering, 25*, 53–62. https://doi.org/10.5755/J01.SACE.25.2.22263

Mayencourt, P., & Mueller, C. (2019). Structural optimization of cross-laminated timber panels in one-way bending. *Structures, 18*, 48–59. https://doi.org/10.1016/J.ISTRUC.2018.12.009

Monahan, J., & Powell, J. C. (2011). An embodied carbon and energy analysis of modern methods of construction in housing: A case study using a lifecycle assessment framework. *Energy and Buildings, 43*, 179–188. https://doi.org/10.1016/J.ENBUILD.2010.09.005

Monteiro, H., Fernández, J. E., & Freire, F. (2016). Comparative life-cycle energy analysis of a new and an existing house: The significance of occupant's habits, building systems and embodied energy. *Sustainable Cities and Society, 26*, 507–518. https://doi.org/10.1016/J.SCS.2016.06.002

Murali Mohan, M., Venugopal Goud, E., Deva Kumar, M. L. S., Kumar, V., Kumar, M., & Dinbandhu. (2021). Parametric optimization and evaluation of machining performance for aluminium-based hybrid composite using utility-Taguchi approach. In *Lecture notes in mechanical engineering* (pp. 289–300). Springer. https://doi.org/10.1007/978-981-16-3033-0_27

Pei, S., van de Lindt, J. W., & Popovski, M. (2013). Approximate R-factor for cross-laminated timber walls in multistory buildings. *Journal of Architectural Engineering, 19*, 245–255. https://doi.org/10.1061/(ASCE)AE.1943-5568.0000117

Pei, S., van de Lindt, J. W., Popovski, M., Berman, J. W., Dolan, J. D., Ricles, J., Sause, R., Blomgren, H., & Rammer, D. R.: Cross-laminated timber for seismic regions: Progress and challenges for research and implementation. *Journal of Structural Engineering, 142*. https://doi.org/10.1061/(ASCE)ST.1943-541X.0001192

Popovski, M. et al. (2012). Seismic behaviour of cross-laminated timber structures. researchgate.net.

Robertson, A. B., Lam, F. C. F., & Cole, R. J.: A comparative cradle-to-gate life cycle assessment of mid-rise office building construction alternatives: Laminated timber or reinforced concrete. *Buildings, 2*, 245–270. https://doi.org/10.3390/BUILDINGS2030245

Santamouris, M., & Vasilakopoulou, K. (2021). Present and future energy consumption of buildings: Challenges and opportunities towards decarbonisation. *e-Prime, 1*, 100002. https://doi.org/10.1016/J.PRIME.2021.100002

Schenk, M., Hübner, C., & Cabrero, J. M.: Cross-laminated timber: A survey on design methods and concepts in practice. *CivilEng, 3*, 610–629. https://doi.org/10.3390/CIVILENG3030036.

Schneider, J., Karacabeyli, E., Popovski, M., Stiemer, S. F., & Tesfamariam, S. (2014). Damage assessment of connections used in cross-laminated timber subject to cyclic loads. *Journal of Performance of Constructed Facilities, 28*. https://doi.org/10.1061/(ASCE)CF.1943-5509.0000528

Shukla, A., Tiwari, G. N., & Sodha, M. S. (2009). Embodied energy analysis of adobe house. *Renewable Energy, 34*, 755–761. https://doi.org/10.1016/J.RENENE.2008.04.002

Simonen, K., Rodriguez, B. X., & De Wolf, C. (2017). Benchmarking the embodied carbon of buildings. *Technology Architecture Design, 1*, 208–218. https://doi.org/10.1080/24751448.2017.1354623

Singh Ahluwalia, M., & Patel, U. (2021). Getting to net zero: An approach for India at CoP-26.

Smyth, M. (2018). A study of the viability of cross laminated timber for residential construction.

Snazzymaps. (2022). Carona: Goa as read on 20th June 2023.

Sudhirsastry, Y. B., Krishna, Y., & Budarapu, P. R. (2015). Parametric studies on buckling of thin walled channel beams. *Computational Materials Science, 96*, 416–424. https://doi.org/10.1016/J.COMMATSCI.2014.07.058

Svortevik, V. J., Engevik, M. B., & Kraniotis, D. (2020). Use of cross laminated timber (CLT) in industrial buildings in Nordic climate—A case study. *IOP Conference Series Earth and Environmental Science, 410*, 012082. https://doi.org/10.1088/1755-1315/410/1/012082

Wang, Z., & Yin, T. (2022). Cross-laminated timber: A review on its characteristics and an introduction to Chinese practices. In *Engineered wood products for construction*. IntechOpen. https://doi.org/10.5772/intechopen.98956

Yadav, S., Sharma, P., Yamasani, P., Minaev, S., & Kumar, S. (2014). A prototype micro-thermoelectric power generator for micro- electromechanical systems. *Applied Physics Letters, 104* (2014). https://doi.org/10.1063/1.4870260/24501

Younis, A., & Dodoo, A. (2022). Cross-laminated timber for building construction: A life-cycle-assessment overview. *Journal of Build Engineering, 52*, 104482. https://doi.org/10.1016/J.JOBE.2022.104482

Zhang, C., Yin, Y., Yan, H., Zhu, S., Li, B., Hou, X., & Yang, Y. (2023a). Centrifuge modeling of multi-row stabilizing piles reinforced reservoir landslide with different row spacings. *Landslides, 20*, 559–577 (2023). https://doi.org/10.1007/S10346-022-01994-5/METRICS

Zhang, W., Kang, S., Liu, X., Lin, B., & Huang, Y. (2023b). Experimental study of a composite beam externally bonded with a carbon fiber-reinforced plastic plate. *Journal of Building Engineering, 71*, 106522. https://doi.org/10.1016/J.JOBE.2023.106522

Zhu, G., Liu, Y., Wang, L., Sang, L., Zhao, K., Zhang, Z., Lin, X., & Qiu, D. (2023). The isotopes of precipitation have climate change signal in arid Central Asia. *Global and Planetary Change, 225*, 104103. https://doi.org/10.1016/j.gloplacha.2023.104103

Redefining Sustainable Performance: Life Cycle Cost and CO_2 Footprint Analyses of Two Distinct Affordable Green Homes Located in India

E. V. S. Kiran Kumar Donthu, Sanket Puranik, Kiriti Sahoo, and Minni Sastry

Abstract Building sustainability in India is a complex and challenging task, but it is also a pressing need. India is a fast-developing country with a huge and growing population. This fast growth has put a strain on the nation's natural resources and the environment. The construction and operation of buildings in India have an enormous impact on energy consumption and carbon emissions. Understanding the energy footprint and life cycle cost of these structures is essential for sustainable development and climate action. This study presents case studies of two energy-efficient and affordable residential units in two cities with distinct climatic conditions. The approach used in this study comprises obtaining the bill of quantities (BOQ) and construction material inventory from the designers and contractors of the buildings. The two case studies are analysed for their carbon footprint and life cycle analysis and compared their sustainable performance with the conventional buildings. It is found that Building 1, an independent home constructed with CSEB, in a Temperate

E. V. S. K. K. Donthu (✉) · S. Puranik · K. Sahoo · M. Sastry
The Energy and Resources Institute (TERI), SRC, Bengaluru, India
e-mail: kirankumar.devs@gmail.com

S. Puranik
e-mail: sanketpuranik@gmail.com

K. Sahoo
e-mail: kiritisahoo@nitc.ac.in

M. Sastry
e-mail: minnisastry@gmail.com

E. V. S. K. K. Donthu
Indian Institute of Technology (BHU), Varanasi, India

S. Puranik
4Kongsberg Digital, Oslo, Norway

K. Sahoo
National Institute of Technology, Calicut, India

M. Sastry
R V College of Architecture, Bengaluru, India

J. Vijayalaxmi (ed.), *Sustainable Building Performance*,
https://doi.org/10.1007/978-981-96-2428-7_5

climate has over 78% reduction in the carbon footprint with over 50% life cycle cost compared to conventional building; Whereas, Building 2, a multi-family housing constructed using monolithic RC wall, in a Warm and humid climate has over 31% reduction in the carbon footprint and over 33% reduction life cycle cost compared to conventional buildings. In addition, it is observed that although their life cycle costs are comparable, the carbon footprint of Building 2 is over five times higher than Building 1. This is majorly due to the level of mechanisation and the embodied energy of the materials used in the construction. Building 2, a multi-family home, has a development density that is more than three times that of Building 1, an independent, detached home. As analysed and critically observed in this study, it is important to consider the energy, environmental, and economic factors while defining sustainable performance considering the building's life cycle.

Keywords Life cycle cost · Affordability · Carbon footprint · Green Home · Sustainability

1 Introduction

In India, the residential sector accounts for over 30–40% of overall energy consumption, making it one of the major consumers of electricity (Mospi, 2021). Building designs and construction methods need to change to take account of the rising global concern over climate change. In 15 years between 2021 and 2036, India's housing demand is growing exponentially at a rate of 310%. There is a good opportunity for achieving energy savings through buildings. The country needs buildings that are energy efficient. There is also a housing shortage in India as well, especially with the LIG and EWS section of society (ICRIER, 2020). Therefore, there is a need for affordable, environmentally friendly housing for sustained growth. There are several debates and discussions on whether energy efficiency and affordability are possible to achieve together. Residential energy consumption reduction using passive technologies has a great potential for achieving CO_2 reduction. Energy-efficient buildings reduce electricity demand by more than 30%, reducing overall energy demand and helping minimise the need for setting up new power plants and reducing associated emissions. About 60% of the basic cost of residential construction is related to building materials, of which a significant portion is made up of cement (18%), bricks (17%), lumber (13%), and steel (10%). Basic construction materials such as bricks, mortar, plaster, and concrete, are produced using a significant amount of the embodied energy. Glass, cement, and plaster boards have primary energy requirements ranging from 60 to 10 GJ/ton, while the production of aluminium, copper, stainless steel, and plastics uses the most energy (primary energy requirements range from 250 to 100 GJ/ton). According to estimates, a building's energy embodied might be equivalent to 15 to 20 years of its energy use while in operation. The construction material quantities for two distinct green buildings are extensively studied in this chapter along with their cost and energy footprint (including operational and embodied energy).

2 Methodology

This study includes case studies for two types of residential units in two cities with distinct climatic conditions that were designed to be energy-efficient and affordable. The methodology used in this study includes obtaining the bill of quantities (BOQ) and construction material inventory from the designers and construction contractors of the buildings. The carbon footprint of the buildings and their life cycle costs, including initial construction and operating costs, are also determined by carefully analysing this data. Based on the review of existing data, the intensity of embodied energy has been referred to from secondary resources for evaluating the building's carbon footprint. The building type, basic construction features, and climate location are described in the following section. The energy bills of the respective house or residential unit have been used to get the operational energy data. Energy-efficient and affordability concepts implemented in the house are described in detail in the following sections. Table 1 presents an overview of the project details.

2.1 Building 1—Independent CSEB House in Bangalore

The first building is an independent Compressed Earth Stabilised Blocks (CSEB) house which is located in Mariyanapallya, Bangalore. The residence is a two-storey (Ground + 1) load-bearing construction. The materials and techniques used for construction of the house are majorly alternative and locally sourced thus lowering the overall cost and embodied energy of the house. CESB used for the house are manufactured using soil from the same site. The roof is of jack arch type construction; profile of the jack arch roof gives lateral or transverse strength to the panels for distributing the load through compressive forces. The strength of a jack arch roof is comparable to the conventional RCC roof due to its structure. An interesting construction technique used in the project is U-shaped mud blocks to make the lintels. Instead of providing a continuous lintel, the architects came up with this cost-effective technique to save material. Doors used in the construction are mostly flush doors. Terracota Jalis were used near semi-open areas. In most parts of the house, clay tiles were used for flooring, which contributes to the overall low embodied energy and costs of the building. Kota stone has been used for kitchen platforms and granite slabs for the shelves.

2.1.1 Passive and Energy-Efficient Features-Building 1

The architecture of the house construction incorporates a few passive energy techniques. A skylight is provided above the stairs lighting up the staircase and the common room on the first floor. A good number of ventilators in every room keep the spaces well-circulated. Windows and ventilators integrated with the roof ensure

Table 1 Summary of the two residential buildings presented in this study

Details about the project		

Climate (as per NBC)	Moderate	Warm and Humid
Location	Mariyanapallya, Bangalore (13° 03′ N, 77° 36′ E)	Oragadam, Chennai (12° 45′ N, 80° E)
Site area	158 m^2	8417 m^2
Built-up area	125 m^2	10,952 m^2 (each apartment unit 74 m^2)
FSI	0.8	1.3
Type of project	Individual (3BHK)	2BHK apartment (Stilt + 4 floors)
No. of units	1	148
Density	26 families/acre	73 families/acre
Distance from city	~ 15 KM	~ 50 km
Construction period	10 months	18 months
Level of mechanisation	Low	High
Skill level of labour	High	Medium
Total cost of construction	₹. 2300/Sq.ft	~ ₹. 1400/Sq.ft
Total life cycle cost (50 years)	₹. 4100/Sq.ft	~ ₹. 4550/Sq.ft
Total carbon footprint	~ 22 kg of CO_2 Eq/m^2/Year	~ 115 kg of CO_2 Eq/m^2/Year

naturally lit and thermally comfortable spaces, thus reducing the load on artificial lighting and ventilation. The CSEB wall construction keeps the interiors thermally comfortable and the ventilators to ensure fresh air, thus lowering the need to switch on fans, especially in winter.

Besides these passive features, energy-efficient fixtures and appliances as well as renewable energy systems have been integrated into the house. Majority of the energy used in the house is sourced from the alternative energy sources like solar PV and solar thermal systems. In terms of internal lighting, only CFLs and LED lights have been used in the entire house. Although the initial cost of these fixtures is quite high the operational energy consumption is comparable. Solar PV system of capacity 1.25 kWp integrated with power conditioning unit (PCU) has been placed on the roof which meets the energy requirements of the house. The house meets its requirement from this renewable source till the battery level is above 20%. When it drops below 20%, the house draws electricity from the grid. When and if the grid fails, the house uses 20% reserve of the solar batteries. A solar water heater with a capacity of 200 L is installed on the roof to suffice the hot water requirement of the house. Although there is a geyser of 2 kW capacity (25 l), it is very seldom used in the house. The total consumption of the house 28 kWh/m^2/annum, whereas the total energy consumption of a conventional house in Bangalore could be 85 kWh/m^2/annum. Currently, 44% of the house's annual energy demand is met by Solar PV and 28% is met by the grid power. About 28% of annual energy is saved due to solar hot water systems installed in the house.

2.1.2 Cost and Embodied Energy Aspects—Building 1

The CESB is beneficial due to its uniform building component sizes, usage of local materials and reduction in transportation energy. Because of uniform surface texture and skilled work, the masonry of the CESB does not require any plastering or painting, thus resulting in material cost savings. The CSEB block is one of the energy-efficient among the alternative materials for walling, consuming only one-fourth of the energy of common burnt clay brick. Though the construction of the jack arch roof in this project is labour intensive, this technique is highly cost-effective. The adoption of this technique helps the users save on a lot of construction material (concrete and steel) and in turn money. Jack arch roof has low embodied energy as compared to the RCC roof because of its quantity of cement/cement per unit volume of the roof. It consists of 40% less steel, however the overall energy content is determined by the embodied energy of the burnt clay tiles used in the roof construction. The roof has better thermal performance because of reduced thermal gains possible due to thermal provided by the brick tiles. U-shaped hollow mud block technique for lintels saves a significant amount of concrete and steel and results in cost-effective and less energy intensive. Flush doors used in this building are less expensive but durable. Terracotta jalis used in this building eliminates the use of steel and results in reduction of the overall costs and embodied energy of the building. Clay floor tiles and Kota stones used in a major part of the building are procured locally, thus resulting in lower embodied energy. Component-wise embodied energy and costs of major items are compared with the conventional construction and the details are tabulated in Appendix 1.

2.2 *Building 2—Multi-Family RCC Housing in Chennai*

The multi-family Reinforced Cement Concrete (RCC) housing is in Oragadam, Chennai. It has implemented several green design features and construction technologies. The building comprises of 2BHK apartments (148 no's) with floor area of 792 sqft each, spread over 2 acres of site. The primary construction material used in this project is cast in situ reinforced concrete (RC) walls (150 mm thick). M20 grade ready mix concrete (RMC) and Fe500 TMT steel reinforcement were used for constructing the RC wall. The project has been certified by IFCs Excellence in Design for Greater Efficiencies (EDGE) Program for its green building design.

2.2.1 Passive and Energy-Efficient Features—Building 2

The site plan of the project is developed with almost 70% of the area as pervious with more green area and less paved area. Enough space has been considered between the apartment blocks which enables mutual shading and free air movement thus reducing the heat gains. Window shading has been recessed with the wall and the interior part of the wall is used as a wardrobe. Simple three-bay units were designed for natural daylight and free flow of air in the living spaces, considering the climate of the city. Full-height windows (floor to lintel) in the living and bedrooms also enhance the ventilation requirement given in the bye-law (20% of the Floor area). These windows, having a clear glazing and optimised shading spread a natural and glare-free lighting inside the living spaces. Ventilators provided above the main door in the living room aid cross ventilation for improved thermal comfort.

Energy efficiency and resource utilisation is one of the key aspects of this project. Each apartment is connected to a fixed load of 3 kW, which prevents the wastage of electricity in the entire housing. Electrical sockets are inbuilt in the R C wall based on the requirement and doesn't allow any additional plug points to capture more loads. Artificial lighting design has been done very optimally, for a uniform lighting distribution. Solar water heating systems have been provided for the top two floors. Standalone solar-based LED lighting has been provided for outdoor lighting which makes the site lighting efficient and reduces the overall dependency on the grid electricity. Solar hot water system reduces almost half of the energy that goes for hot water in this project. Overall about 10 kWh/m^2/year estimated savings would be possible with this system. This makes the project energy-efficient upto 74 kWh/m^2/year, whereas the energy consumption of a conventional house in Chennai would be 100 kWh/m^2/year.

2.2.2 Cost and Embodied Aspects—Building 2

Spaces inside the house were designed to comply with the standard size requirements of the city for an effective use of floor space. Similarly, the corridor widths were also kept to the minimum of bye-law requirement valuing the real estate cost and making the project affordable. Simple design has been used for windows to reduce the cost of the same. Recycled UPVC door/window framed clear glazing has been used which contributes to the overall energy footprint reduction of the building. Composite doors have been used to reduce the overall energy footprint and cost of the project. There are other key features like granite kitchen platforms and its fixing using simple L-brackets makes the building cost-effective. Granite being a local material results in lower embodied energy. Kota stone has been used for staircases. Component-wise embodied energy and costs of major items are compared with the conventional construction and the details are tabulated in the Annexure.

3 Analysis

3.1 *Life Cycle Cost and Carbon Footprint—Building 1*

Energy efficiency and affordability of Building 1 was analysed by calculating carbon footprint and Operational and Maintenance (O&M) Costs during the building's lifetime. Energy and water consumption of Building 1 and its dependency on the grid and external supply was significantly low. Cost of electricity was assumed as ₹ 5.3/kWh and for water as ₹ 100 per 1000 l. The escalation rate for all three costs (O&M, Energy and Water) was assumed to be 2% (fixed) every year. This analysis doesn't include the appreciation/depreciation costs and taxes on the built up property. Table 2 presents the costs of studied case and the conventional case based on these assumptions. The analysis also presents the payback for Building 1 and it can be noticed that the higher initial cost spent on the building pays back in 2.5 years. The initial cost of construction of Building 1 was estimated as ₹ 2345/sq.ft whereas the conventional house would be ₹ 2230/sq.ft. Assuming 50 years life of the building, it was observed that the initial investment for Building 1 was about 5% higher compared to a conventional house.

Table 2 Energy and carbon footprint of Building 1 and conventional house

		Building 1	Conventional house
Total embodied energy (kWh)	Civil structure	84,715	119,031
	Solar PV (1.25 kWp)	10,500	0
	Solar water heater (200 l)	4854	0
Total operational energy taken from grid in 50 years lifetime (kWh)		43,750	531,250
Total energy footprint in 50 years lifetime (kWh)		143,819	650,281
Total carbon footprint in 50 years lifetime (tons of CO_2 equivalent)		136	614

Table 3 Initial and operational costs of Building 1 and conventional house

		Building 1	Conventional house
Total initial costs (₹)	Building	2,592,682	28,34,479
	Energy efficient features (like SPV, eff. lighting and SWH)	279,663	6153
	Water-efficient features (Like STP, eff. fixtures)	281,110	94,600
Total operational costs including purchase of electricity and water for 50 years (₹)		2,363,539	8,227,977
Total life cycle costs for 50 years (₹)		5,516,994	11,163,208
Total cycle costs for 50 years (₹/ft^2)		4102	8300

To calculate total energy footprint and resultant carbon footprint of the building both initial energy of the structure (embodied energy) and the operational energy (electricity used) during its lifetime were considered. Operational energy was considered from the actual data available in the electricity bills. Energy that was spent on transporting materials to site, handling the construction equipment and preparation of materials was neglected in the analysis. Assuming all these energies in terms of electricity (kWh), carbon footprint was calculated for Indian conditions. The total embodied energy of building 1 was estimated to be 2.44 GJ/m^2 whereas the conventional load-bearing construction would be 3.42 GJ/m^2. The operational energy taken from the grid of Building 1 was significantly low; this is because of the energy-efficient and renewable energy features implemented in the house. Embodied energies of solar PV and solar thermal systems for hot water systems used in the house were derived from the earlier studies. The carbon footprint of Building 1 was estimated as 22 kg of CO_2 Eq/m^2 built-up area/year and the conventional house used for comparison would be 98kgs of CO_2 Eq/m^2 built-up area/year. These figures were compared with the conventional building using the data compiled in Table 3.

From both the cost and energy footprint analyses it was observed that the studied case (Building 1) has about 78% lower carbon footprint and is about 50% more affordable compared to conventional case. The carbon footprint of Building 1 was observed as 1.1 Tons of CO_2/m^2 built up area whereas for the conventional house it was 4.9 Tons of CO_2/m^2 built up area. The initial and O&M costs of Building 1 in its entire life cycle were observed as 4102 Rs./Sq.ft of built up area whereas for a conventional house, it was 8300 Rs./Sq.ft. Figure 1 presents the incremental energy and costs plotted for 50 years of lifetime.

3.2 *Life Cycle Cost and Carbon Footprint—Building 2*

The initial cost for Building 2 was about 8% higher due to additional energy-efficient and water-efficient features implemented in the project. The operational energy and

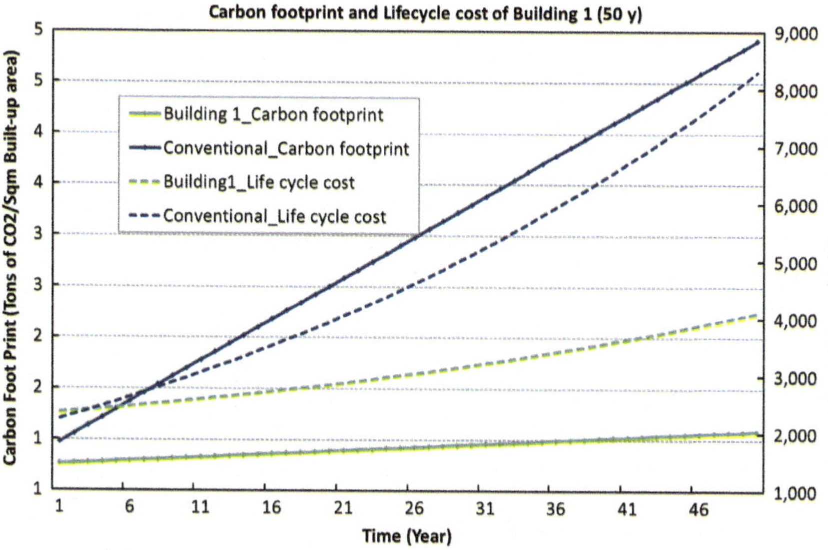

Fig. 1 Life cycle cost and carbon footprint of Building 1 (50 Y)

water costs were taken from the actual data taken from the electricity and water bills. Energy and water consumption of Building 2 and its dependency on the grid and external supply was low. The operational energy taken from the grid of Building 2 was low as presented in the previous section; this is because of the energy-efficient and renewable energy features implemented in the project. Since the energy saved by Solar PV-based LED streetlights is very minimal, it was not accounted into the calculations. Table 4 presents the costs of Building 2 and the conventional case based on these assumptions. The analysis also presents the payback for Building 2 and it can be noticed that the higher initial cost spent on the building pays back in 2 years. The initial and O&M costs of Building 2 in its entire life cycle was observed as ₹4950/Sq.ft of built up area whereas for a conventional house, it was ₹7450/Sq.ft.

The total embodied energy of Building 2 was assessed as 8.47 GJ/m^2 built-up area whereas the conventionally framed construction with cement concrete block

Table 4 Energy and carbon footprint of Building 2 and conventional house

		Building 2	Conventional house
Total embodied energy (kWh)	Civil structure	25,753,853	9,851,555
	Solar water heater (7500 l)	182,041	0
Total operational energy taken from grid in 50 years lifetime (kWh)		40,522,400	54,760,000
Total energy footprint in 50 years (kWh)		66,458,294	64,611,555
Total carbon footprint in 50 years (tons of CO_2 equivalent)		62,737	90,993

masonry would be 3.26 GJ/m^2 built-up area. The recyclability aspect of aluminium formwork in case of Building 2 was not into taken account. Thus overall, the carbon footprint of Building 2 was estimated as 115 kg of CO_2 Eq/m^2 built-up area/year and the conventional house used for comparison would be 112 kg of CO_2 Eq/m^2 built-up area/year (Table 5).

From both the cost and energy footprint analyses it was observed that there was not much difference in carbon footprint of Building 2 and the conventional case. About 33% affordable compared to conventional case. The total carbon footprint was observed as 6 tons of CO_2/m^2 built up area for both Building 2 and the conventional case (Fig. 2).

Table 5 Initial and operational costs of Building 2 and conventional house

		Building 2	Conventional house
Total initial costs (₹)	Building	131,136,882	121,746,566
	Energy-efficient features (like eff. lighting and SWH)	374,400	1,153,000
	Water-efficient features (like STP, eff. fixtures)	5,704,510	1,515,492
Total operational costs including purchase of electricity and water for 50 years (₹)		446,795,003	753,404,487
Total life cycle costs for 50 years (₹)		584,010,796	877,819,545
Total life cycle costs for 50 years ₹/Sq.ft)		4954	7446

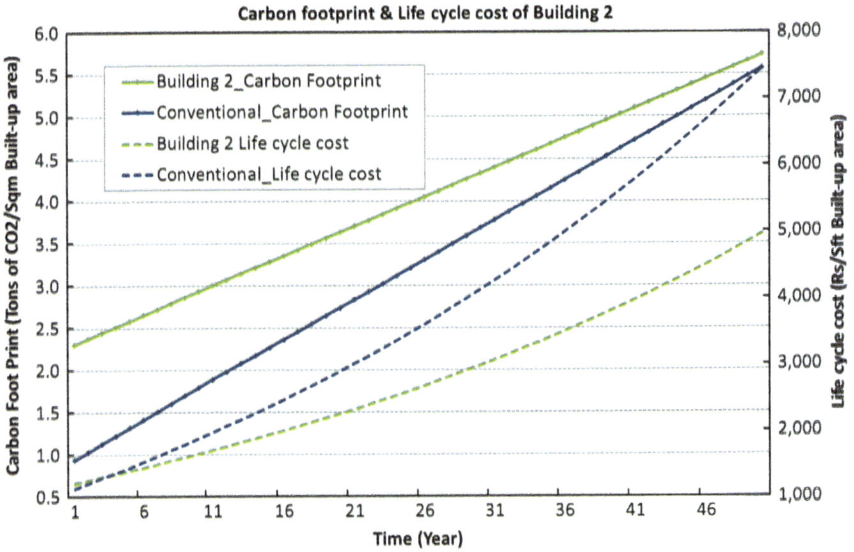

Fig. 2 Life cycle cost and carbon footprint of Building 2 (50 Y)

4 Discussion

The above two sections present data and analysis of two distinct energy-efficient and affordable residential buildings located in two Indian cities namely Bangalore and Chennai which fall under Temperate and Warm-Humid climate zones respectively. Table 1 summarises the building carbon footprint and life cycle cost of the buildings. Although their life cycle costs are comparable, the carbon footprint of Building 2 is over 5 times higher than Building 1. This is majorly due to the level of mechanisation and the embodied energy of the materials used in the construction. Mechanisation of construction of Building 2, in terms of on-site assembly, replicable and consistent production of building components have contributed to control of timelines and overall costs of the project. Prefabricated aluminium formwork was used to cast reinforced concrete walls and reduce the wastage and reworks, and thus results in a better quality of construction. The construction of this housing project was completed in 18 months of duration due to various alternative construction technologies implemented on-site. The monolithic concrete shear walls constructed in this project also meet the seismic code compliance requirements for Chennai. Based on the preliminary data received from the developer, although the initial investment in aluminium formwork is high, its reusability (number of times it can be used) is significantly high compared to a conventional wooden formwork. The labour cost is also almost half of the cost it takes for a conventional construction. The most important parameter is the time of construction using this technology which makes the construction 10 times faster than a conventional construction. The formwork was sufficient for constructing each apartment unit at a time. When compared to cost of a concrete block masonry, it was observed that the cost of RC wall construction becomes equal at 115th repetition and affordable thereafter.

When presenting these scenarios, a question comes up regarding the buildings' sustainability. Building 2's energy footprint, measured in terms of the embodied energy of the construction material, is over five times more than Building 1's, even though the life cycle cost is not much different. Because the residents of the different homes find both construction technologies acceptable, this study does not address the social elements of the three main pillars of sustainability. The carbon footprint and other environmental factors, however, have a significant impact. The number of families per acre, or the development density, is yet another difference. Building 2, a multi-family home, has a density that is more than three times that of Building 1, a standalone detached home. Therefore, it is crucial to consider the energy, environmental, and economic factors covered in this study when defining sustainable performance throughout the building's life cycle.

5 Conclusion

This chapter provides a detailed analysis of two different green and affordable buildings located in two southern Indian cities. The analysis is based on actual data from construction costs and electricity bills, considering a 50-year lifespan for the buildings. It can be observed from the study that Building 1, an independent home has over 78% reduction in the carbon footprint with over 50% life cycle cost compared to conventional buildings; Whereas, Building 2, a multi-family housing has over 31% reduction in the carbon footprint and over 33% reduction life cycle cost compared to conventional buildings. We can observe that these two energy-efficient buildings, despite having higher initial embodied energy due to energy-saving equipment, have a significantly lower overall carbon footprint over their lifetime because of reduced operational energy. Similarly, some construction methods can be cost-effective through material reuse, but the choice of materials significantly impacts the building's total carbon footprint. Therefore, while studying buildings' energy, environmental, and financial performance, it is crucial to highlight the holistic dimensions of sustainability.

Annexure 1

Component-wise costs of the Building 1 when compared to conventional construction

S. No.	Building component	Description	Total cost (in Rs.)	Description	Total cost (in Rs.)
1	Site works	Excavation, termite and water proofing and filling	33,111	Excavation, termite and water proofing and filling	33,111
2	Foundation	Stone masonry	196,998	Stone masonry	196,998
3	Plinth	RCC	52,452	RCC	52,452
4	Beams	RCC	35,614	RCC	35,614
5	Lintel	U-block lintel	29,275	RCC	52,452
6	Walls	Stabilised mud blocks—9″ thick and 4″ thick	373,768	Burnt clay brick—8″ thick and 4″ thick	448,522
7	Plastering	Cement, mud plastering	162,898	Cement plastering	245,588
8	Roof	RCC flat roof	36,304	RCC	413,075
		Jack arch roof	327,627		
9	Flooring	Vitrified and clay tiles	99,812	Terrazzo flooring	38,226
10	Windows	Glazing and shutters with grill	188,084	Glazing and shutters with grill	188,084

(continued)

(continued)

S. No.	Building component	Description	Total cost (in Rs.)	Description	Total cost (in Rs.)
11	Window shade	Chajja and Loft	14,004	Chajja and Loft	14,004
12	Doors	Flush doors	142,916	Flush doors	142,916
13	Staircase	Folded plate RCC staircase	7411	Folded plate RCC staircase	7411
14	Handrail	Teak wood	11,762	Teak wood	11,762
15	Cupboards	Granite slabs (part of civil structure)	269,560	Granite slabs (part of civil structure)	269,560
16	Painting	Only internal	97,106	Both external and internal	194,213
17	Additional features	Skylight (civil works)	23,487	Nil	0
18	Services (electrical)	Excluding energy-efficient fixtures	301,084	Excluding energy-efficient fixtures	301,084
19	Services (plumbing)	Excluding rainwater components	189,408	Excluding rainwater components	189,408
20	Water-efficient fixtures	Domestic and sanitary	103,630	Conventional	64,769
21	Rainwater harvesting system	RW filter, Sump tank construction (10 K L capacity), water level controller, pipes	94,600	RW filter, Sump tank construction (10 K L capacity), water level controller, pipes	94,600
22	Waste water treatment system	Reed bed system	81,433	Nil	0
23	Energy-efficient lighting fixtures	LEDs and CFLs	24,610	Conventional (incandescent and FTL)	6153
24	Solar hot water system	Solar thermal system (125 l)	38,000	Nil	0
25	Photovoltaic panels	1.25 kWp capacity (including inverter, batteries, cabling etc..,)	218,500	Nil	0
		Total cost (Rs.)	3,153,455	Total cost (Rs)	3,000,000
		Total cost (Rs./Sq.ft)	2345	Total cost (Rs/Sq.ft)	2230

Component-wise embodied energy of Building 1 when compared to conventional construction

S. No.	Building component	Description of the material	Total EE of component (MJ)	Description of the material	Total EE of component (MJ)
1	Foundation	Stone masonry	73,060	Stone masonry	73,060
2	Plinth	RCC	15,628	RCC	15,628
3	Beams	RCC	14,382	RCC	14,382
4	Walls	Soil–cement block masonry (7% cement)	46,678	Burnt clay brick masonry	137,238
5	Roof	RCC flat roof	30,001	RCC flat roof	141,318
		Jack arched roof	85,394		
6	Flooring	Clay tiles	5070	Terrazzo flooring	7582
		Vitrified tiles	7158		
7	Lintel	RCC filled in U-blocks	4698	RCC	15,628
8	Window shades-chajja	RCC	10,552	RCC	10,552
9	Doors	Teak door	194	Flush doors	8483
		Flush doors	7519		
10	RCC staircase		4641	RCC staircase	4641
	Total embodied energy (MJ)		304,975	Total Embodied Energy (MJ)	428,513
	Total embodied energy (GJ/100 m^2)		244	Total embodied energy (GJ/ 100 m^2)	342

Annexure 2

Component-wise costs of the Building 2 when compared to conventional construction

S. No.	Building component	Description of the material	Total cost (in Rs.)	Description of the material	Total cost (in Rs.)
1	Site works	Cut and fill, termite proofing, Trenching for foundation	254,383	Cut and fill, termite proofing, Trenching for foundation	254,383
2	Scaffolding/ formworks	Aluminium formworks (used for constructing every 350 units)	4,090,887	Wooden formworks (used for constructing every 6 units)	28,636,208

(continued)

(continued)

S. No.	Building component	Description of the material	Total cost (in Rs.)	Description of the material	Total cost (in Rs.)
3	Foundation and plinth	RCC	5,799,296	RCC	5,799,296
4	Columns and beams	RCC	3,948,392	RCC	3,948,392
5	Walls	Reinforced concrete (RC) wall	47,095,300	Cement concrete block Masonry	23,547,650
6	Painting	Painting works	6,221,597	Painting works	6,221,597
7	Roof	RCC	4,515,220	RCC	4,515,220
8	Flooring	Ceramic and vitrified tile	11,572,002	Terrazzo	10,514,800
9	Windows	uPVC framed	6,995,938	Aluminium framed	3,497,969
10	Doors	Flush doors	5,555,394	Flush doors	5,555,394
11	Staircase	RCC	1,776,413	RCC	1,776,413
12	Railing	Steel	779,940	Steel	779,940
13	Waterproofing	Waterproofing works	1,785,270	Waterproofing works	1,785,270
14	Building services—electrical	Excluding efficient fixtures	10,012,936	Excluding efficient fixtures	10,012,936
15	Building services—plumbing	Excluding efficient fixtures	9,575,006	Excluding efficient fixtures	9,575,006
16	Water-efficient fixtures	Low flow faucets, showers and WCs	2,424,788	Conventional	1,515,492
17	Rain water harvesting system	Gutters, pipes, storage tank work, purification system	2,079,722	Nil	0
18	Waste water treatment system	DEWATS system (9000 LPD)	1,200,000	Nil	0
19	Energy-efficient lighting features	CFLs	732,000	Incandescent	183,000
20	Energy-efficient fans	Energy-efficient fans	1,480,000	Conventional	888,000
21	Solar hot water system	Solar flat plate collector (7500LPD)	1,332,000	Nil	0

(continued)

(continued)

S. No.	Building component	Description of the material	Total cost (in Rs.)	Description of the material	Total cost (in Rs.)
22	Solar PV—LED street lighting	14 W LED luminaires (20 Nos)	200,000	150 W Sodium vapour (20 Nos)	82,000
23	Labour cost	At Rs. 60/m^2	710,400	At Rs. 140/m^2	1,657,600
24	Miscellaneous	Lumpsum amount	1,000,000	Lumpsum amount	1,000,000
		Total cost (in Rs.)	131,136,882	Total cost (in Rs.)	121,746,566

Component-wise embodied energy of the Building 2 when compared to conventional construction.

S. No.	Building component	Description	Total embodied energy (MJ)	Description	Total embodied energy (MJ)
1	Scaffolding/ formworks	Aluminium	47,444,400	Wood	46,000
2	Foundation and plinth	RCC	4,174,839	RCC	4,174,839
3	Columns and beams	RCC	3,146,209	RCC	3,146,209
4	Walls	RCC	25,456,000	Solid concrete blocks	14,492,160
5	Plastering	No Plaster	0	Cement	710,400
6	Roof	RCC	2,765,864	RCC	2,765,864
7	Flooring	Ceramic	4,878,000	Terrazzo	2,512,170
8	Windows	UPVC	2,732,720	Aluminium	5,502,120
9	Doors	Flush	856,032	Flush	856,032
10	Staircase	RCC	1,259,807	RCC	1,259,807
	Total embodied energy (MJ)		92,713,871	Total embodied energy (MJ)	35,465,601
	Total embodied energy (GJ/100 m^2 built up area)		843	Total embodied energy (GJ/100 m^2 built up area)	323

References

Energy Statistics India. (2021). Ministry of Statistics and Program Implementation. https://mospi.gov.in/publication/energy-statistics-india-2021. Accessed 01 Nov 2024

Roy, D., Meera, M. L. (2020). *Housing for India's low-income urban households: A demand perspective.* Working paper 402, Indian Council for Research on International Economic Relations (ICRIER). https://icrier.org/pdf/ES/ES_Housing_for_India.pdf. Accessed 01 Nov 2024

Sustainability Through Energy Efficiency of Low-Income Housing-A Form-Based Iterative Design Approach Using ENVI-Met in Warm-Humid Climate

Gandham Kiranjee and J. Vijayalaxmi

Abstract The global climate crisis necessitates urgent action to reduce greenhouse gas emissions and mitigate the impacts of climate change in high-impact sectors like construction. With demand for low-income housing rapidly rising in India, there exists a unique opportunity to integrate energy-efficient design and planning to significantly reduce energy consumption. The development of Net Zero Energy Communities (NZECs) within low-income housing is critical in this context. This study conducted a comprehensive investigation into the design of an NZEC within an existing low-income housing community in the warm-humid climate of Visakhapatnam, India. The research involved an intensive field study focused on qualitative aspects, including spatial attributes, and residents' energy and resource use patterns, which were assessed through Post-Occupancy Evaluation (POE). Quantitative analysis was conducted through simulations, evaluating the performance of the built environment and testing various design iterations. The findings highlight significant shortcomings in the community's baseline design, particularly in terms of thermal regulation and airflow. By adjusting layout, street, and building-level parameters, the optimized design achieved a 50% reduction in ambient temperature at the layout level and enhanced street-level air velocity by 25–30%, creating a more comfortable and energy-efficient environment. This research provides an alternative approach to current low-income housing typologies, offering a viable model for scaling NZECs in social housing projects across India. The study's outcomes contribute to India's sustainability targets and demonstrate the potential of NZECs to meet the housing sector's rising energy demands.

Keywords Energy-efficient communities · Low-income group housing · Post-occupancy evaluation (POE) · Iterative design · ENVI-met

---

G. Kiranjee (✉)
Environmental Design Solutions, New Delhi, India

J. Vijayalaxmi
School of Planning and Architecture, Vijayawada, India

J. Vijayalaxmi (ed.), *Sustainable Building Performance*,
https://doi.org/10.1007/978-981-96-2428-7_6

Abbreviations

PAT Potential Air Temperature
MRT Mean Radiant Temperature
SVF Sky View Factor

1 Introduction

At COP26 under the Glasgow Climate Pact, India pledged to achieve net zero emissions by 2070. To reach this target, India has proposed key strategies that emphasize policy shifts, technological innovation, and sustainable practices, marking India's commitment to a low-carbon future as shown in Fig. 1.

Achieving India's 2070 net zero commitment requires immediate action. Major transformations must begin now across diverse sectors—including fishing, agriculture, packaging, construction, tourism, water and waste management, energy, and wellness and lifestyle. Each of these areas will play a critical role in driving India's sustainable transition. (National statement at CoP 26, source: Ministry of External Affairs, GOI). This also helps in combating the global climate crisis.

The construction industry consumes nearly 35% of the nation's energy and is indirectly responsible for more than 50% of GHG emissions (D'Amico et al., 2021). To meet the net zero commitment, intensive measures at large scale must be implemented. The role of Net zero energy buildings and Net zero energy communities is considered prominent in construction industry which helps in creating sustainable habitats, resource efficiency, and cutting down GHG emissions.

The implementation of the net-zero concept at the community level has been limited due to its complexity and the multi-faceted approach required (Carlisle et al., 2009). The idea that "the whole is greater than the sum of the parts" is particularly true for Net Zero Energy Communities (NZECs), where the collective benefits at the community level are far more impactful than individual efforts (Mavrigiannaki et al., 2021). Neighborhoods and communities offer opportunities for low-impact heating, cooling, and electricity generation. Adopting net-zero strategies at the community or campus scale can also lead to cost efficiencies through shared services (Kallushi

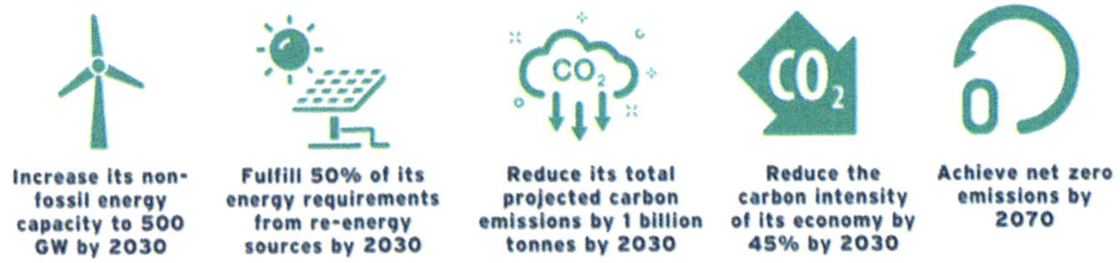

Fig. 1 The nation's targets to be achieved by 2070. *Source* GOI

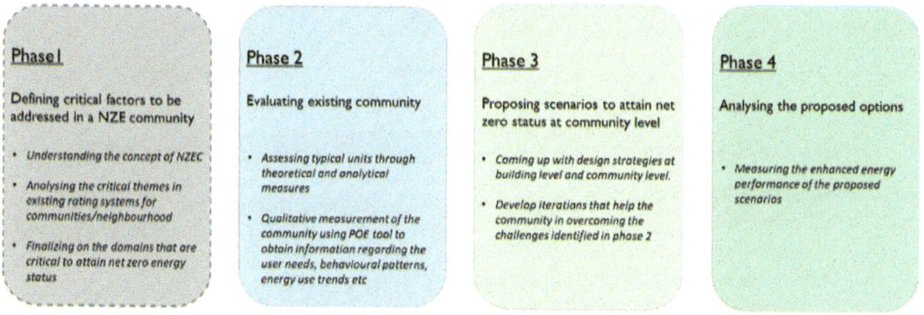

Fig. 2 Research approach taken for the study. *Source*: Authors

et al., 2012). Further, they contribute to attaining SDGs focused on sustainable cities and communities, Health and Well-being, and Responsible Consumption.

National programs like the Pradhan Mantri Awas Yojana-Urban (PMAY-U) have significantly contributed to affordable housing. Since 2015, about 11.8 million dwelling units have been sanctioned under PMAY-U (MoHUA, GoI, 2024). However, India still faces a housing shortage of over 50%, as reported by the Ministry of Housing and Urban Affairs, which has led to the development of numerous projects for Economically Weaker Sections (EWS) and Low-Income Groups (LIG) in urban areas.

While the primary focus of social housing remains affordability, there is increasing emphasis on incorporating sustainable practices, such as the use of eco-friendly materials, renewable energy, and resource efficiency. Although the adoption of net-zero concepts in social housing is still limited, India's goal of achieving net-zero emissions by 2070 highlights the importance of exploring how these communities can contribute to this target. To address this, the study suggests design strategies for rethinking traditional social housing, with the goal of achieving net-zero energy status through a clear and structured approach, as depicted in Fig. 2.

2 Literature Review

A number of definitions have been offered for NZE buildings and, more recently, NZE communities. An NZE community consumes no more energy than it is able to produce annually through renewables located in the community's perimeter or in surrounding non-urban areas (Carlisle et al., 2009). It generally has the following network of critical aspects such as renewable, heating, and cooling as shown in Fig. 3.

From Fig. 4, it is evident that evaluating a community is a subjective task and there is no tailor-made procedure or solution apart from rating systems to quantify the performance of a community. Campos (2020) and Boarin et al. (2018) strongly believe that evaluation systems should consider the interaction of individuals with the built environment. This perspective of Post-Occupancy evaluation is being practiced

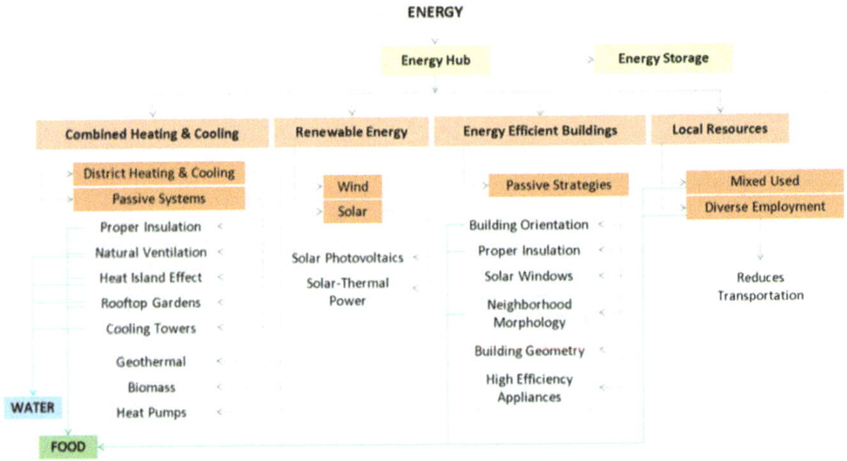

Fig. 3 Critical aspects to be addressed in a NZEC (Franks, 2020)

at building-level energy systems and services but often neglected at a community level.

With the absence of an accepted structured evaluation at the community scale, various studies have adopted a variety of assessing tools to evaluate. Few such models and methodologies that were adopted in recent times are as shown in Table 1.

Through literature, it is found that the rating systems comprehensively evaluated the quantitative aspects of the community to a greater extent when compared to existing theories in literature, but lacked a structured approach to assess the qualitative aspects in the built environment.

Rating system	LEED-ND	BREEAM	CASBEE-UD	Green Star Communities	HQE²R	Pearl community rating system	IGBC Green townships	GRIHA LD
Themes Addressed	• Smart location and linkage • Neighbourhood pattern and design • Green infrastructure and buildings • Regional priority and innovation	• Governance • Social and economical well-being • Resources and energy • Land use and ecology • Transport and movement • Innovation	• Natural environment • District service functions • Contribution to the community • Impact on micro climate, façade and landscape • Social infrastructure • Management of the local environment	• Governance • Design • Liveability • Economic prosperity • Environment • Innovation	• Enhance heritage & conserve resources • Local environmental quality • Diversity • Community integration, Social liveability	• Integrated development process • Natural systems • Liveable communities • Precious water • Resourceful energy • Stewarding materials • Innovative practice	• Site selection & planning • Land use planning • Transportation planning • Infrastructure resource management • Innovation in design and technology	• Carrying capacity and carbon foot print • Site planning • Energy • Water and waste water • Solid waste management • Transport • Social

Fig. 4 Major themes addressed by community rating systems (various rating systems)

Table 1 Methods of analyzing communities by researchers (compiled by Authors)

No	Journal and Author(s)	Approach	Methodology	Findings	Inferences
1	***Anne-Françoise Marique, Sigrid Reiter, (2014)*** A simplified framework to assess the feasibility of zero-energy at the neighborhood/ community scale	Mathematical model	Adopted a simplified framework to assess the feasibility of zero-energy at community scale	Improving the energy efficiency of building stock is one of the crucial factors to achieve net zero	Trial and error method which is applicable to already existing communities with energy consumption data
2	***Sarika Bahadure, Rajashree Kotharkar (2017)*** Framework for measuring sustainability of neighborhoods in Nagpur, India	Quantitative assessment against benchmarks in literature	Twenty sub-indicators are selected under different domains and evaluated for twelve neighborhoods	Neighborhoods with higher percentage of mixed-use and population density showed less levels of sustainability	Data collected through field survey and the results obtained do not give any design guidelines for built environment. However, the optimum % of open areas and roads can be deduced
3	***Boarin P (2018)*** Post-occupancy evaluation of neighborhoods	Qualitative assessment using POE	A new POE framework has been proposed based on the literature study for studying environmental performance of a community	Need for adopting mixed mode approach in data collection to understand user experience and performance of built environment	User perspective is obtained on desirable features in community. But, no information on quantification could be obtained

(continued)

Table 1 (continued)

No	Journal and Author(s)	Approach	Methodology	Findings	Inferences
4	***Salter, J.*, et al. *(2020)*,** Iterative 'what-if' neighborhood simulation: energy and emissions impact	Simulations using Infill design	This is based on profiles from census data, geospatial analysis, and energy data that characterizes neighborhood patterns. Modeling is carried out in Rhino and iterations are compared	Retrofitting case and replacement are analyzed and the study showed nearly 20–25% increase in energy efficiency	Even though the study shows best case scenario through iterations, it does not account to site layout options. The iterations are generated only with the existing masterplan
5	***Mavrigiannaki* A, et al. *(2021)*** Zero energy concept at neighborhood level: a case study analysis	Energy consumption assessment using simulation and on-site monitoring	Study is carried out in Italy. 2 residences of G, G + 1 floors in an NZE neighborhood are selected. Energy systems are monitored before occupancy and year after occupation. The energy systems are simulated to estimate the performance	24% lower investment is observed in the case of NZE at neighborhood than at the building level	This methodology is applicable for assessing communities that are existing and constantly monitored for energy consumption No design inputs can be drawn from this evaluation method

3 Materials and Methods

A Post-Occupancy Evaluation (POE) was conducted to gather user perceptions of the community under study. A non-interventional field survey was administered to a representative sample, using a structured set of 90 questions to collect comprehensive data. The survey covered the following areas:

- General Information: Demographics of respondents and family structure.
- Dwelling-Level Satisfaction: Feedback on layout, dwelling size, and indoor environment quality.
- Comfort Within Units: Assessment of thermal comfort, ventilation, indoor air quality, and daylighting.
- Household Consumption Patterns: Analysis of electricity use, types of appliances, seasonal and usage patterns.
- Fuel and Water Use: Data on cooking fuel consumption, water supply, and usage.

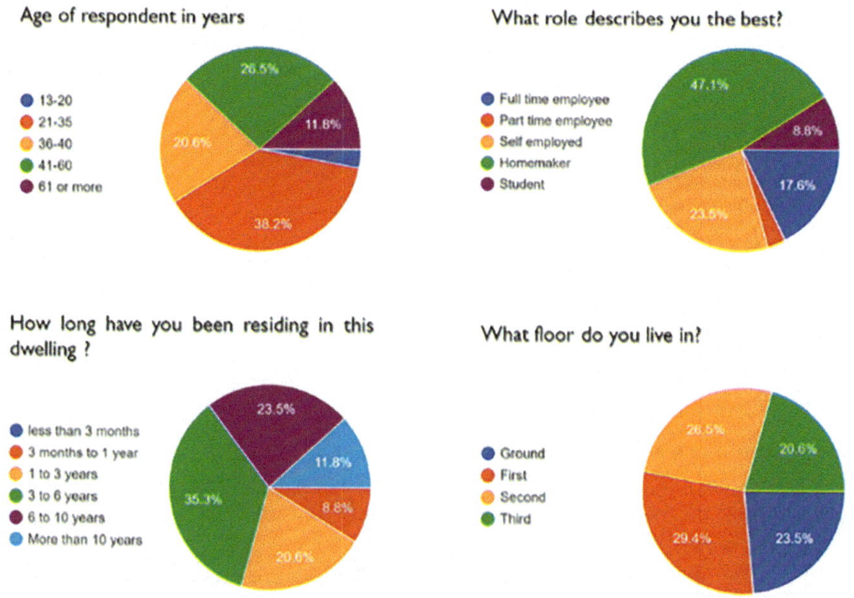

Fig. 5 Respondent profile

- Mobility and Commuting: Information on respondents' commuting patterns and travel to workplaces.

The sample consists of 102 respondents from diverse age groups, representing 236 dwelling units (DU), as illustrated in Fig. 5, which shows their building floor levels and unit orientations. Of these respondents, 62% are women, and 71% have resided in the community for over three years as tenants. Consistent with the methodology found in the literature, a simulation-based iterative design approach was applied for the quantitative aspects of this study.

The quantitative assessment focuses on layout-level parameters, including ambient temperature, wind speeds within clusters, and daylighting in each dwelling unit relative to obstructions and adjacent buildings. ENVI-Met was used for the initial temperature and wind speed analysis, while Velux software supported the daylighting analysis within individual units.

4 Study Area

The study area is located in Marikavalsa, Visakhapatnam (Fig. 6) which falls in the Warm-humid climatic zone as per climate classification of India (NBC and ECBC). Warm-humid climates, characterized by high temperatures and humidity levels, pose significant challenges to energy consumption and building performance.

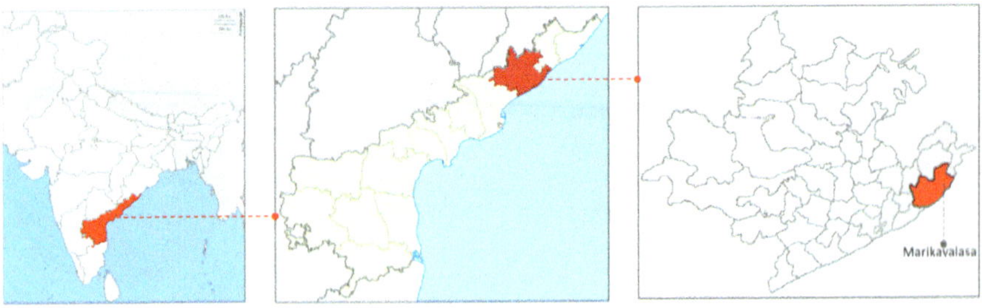

Fig. 6 Location of the study area w.r.t district and state map. *Source* Maps of India

These conditions are prevalent in developing nations of South and Southeast Asia, the Caribbean, and parts of South America, require passive design strategies to mitigate thermal discomfort, reduce energy demand, and enhance sustainability in an affordable way.

It can be observed from Table 2 that the temperature range in every month is considerably similar and there is no major variation in the annual range. Also, the amount of precipitation during a year is considerably high close to 1200 mm.

The average Relative Humidity of Visakhapatnam is around 79%. However, it varies from around 75% during Winter (December) to 83% during the Monsoon (September). The most humid month of the year is September with humidity ranging from 63.7 to 93.9% whereas the least is in the month of December, varying from 44.1 to 99.2% making effective ventilation a prerequisite for comfortable indoors.

The project selected for post-occupancy evaluation is an Economically Weaker Section (EWS) and Low-Income Group (LIG) housing development located in a suburb of Visakhapatnam (Fig. 7). Built prior to 2010 under the Rajiv Gruhakalpa affordable housing scheme, the project aims to address housing needs for the urban poor. Spanning approximately 15 acres, it includes 3200 houses and serves a population of nearly 9400.

For the conceptual design, a representative block was selected, containing 24 units (768 dwelling units) and housing close to 2500 residents. A sample of eight units, or one-third of the total, was surveyed for the study (Fig. 8). The units are

Table 2 Average monthly temperature and precipitation of Visakhapatnam

	Jan	Feb	Mar	April	May	June	July	Aug	Sept	Oct	Nov	Dec
Avg. Temp °C	23.5	25.4	27.9	29.9	31.2	30.9	29.4	29.2	28.8	28	25.8	23.6
Min Temp °C	19.5	22	24.4	26.5	28	27.8	26.6	26.4	26	25	22.4	20
Max Temp °C	27.5	28.9	31.5	33.4	34.5	34.1	32.2	32.1	31.7	31.1	29.2	27.3
Precipitation mm	10	9	12	21	52	106	136	147	177	259	73	6

Source Indianclimate.com—meteorological data archive

Fig. 7 Site boundary and typical block for analysis in red. *Source* Google Earth

clustered in groups of eight, with a Ground plus three floors (G + 3) configuration, and are oriented along an east–west axis. The streets, which are 9 m wide, follow the same layout, and most of the vegetation is situated along the southern side of the unit clusters. Minimal vegetation is observed at the household level, with sporadic greenery along the roadside. Shading analysis results are depicted in Fig. 9.

Each unit's typical plan is shown in Fig. 10. The unit has a total building height of 12 m. The unit has 2 staircases with 4 dwelling units arranged on either side of staircase. Size of each dwelling unit is 30 m^2 comprising of a living room, bedroom, kitchen, and a bathroom of areas 12.4, 8.4, 4.0, 2.8 m^2, respectively. The exterior units have bedrooms opening to adjacent buildings and the interior units share a common wall. This typology is typical across many such low-cost housing projects across the country.

5 Survey Results Related to Spatial Attributes of the Dwelling Unit

The space and comfort survey indicates a notable demand for additional living space and aesthetics; however, there is an even stronger preference for habitable spaces with effective ventilation (Fig. 11). To understand residential energy consumption patterns, a household survey was conducted, gathering data on appliance types, usage intensity, and monthly electricity expenses. Comparing the Energy Performance Index (EPI) of all 32 dwelling units within a block to a baseline EPI reveals that 18% of the units have an EPI exceeding the baseline. Notably, 50–60% of units show significantly lower EPI values, ranging from 10 to 15 kWh/m^2/year, as shown in Fig. 12.

Fig. 8 Schematic design of the housing (Baseline)

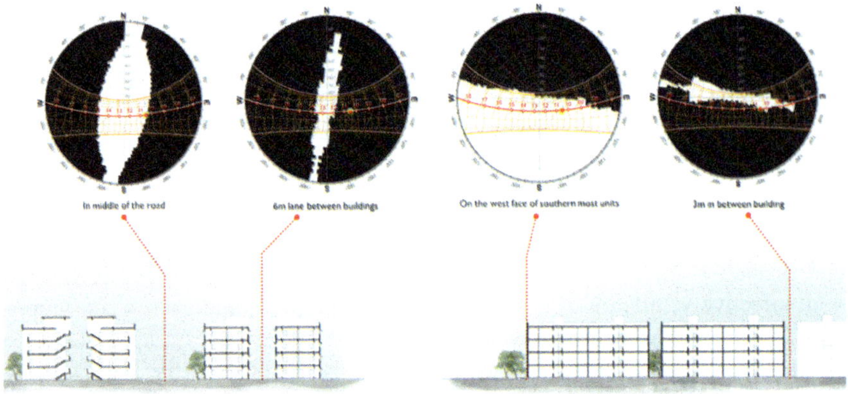

Fig. 9 Shading observed at respective points at section A-A and B-B for the Base case

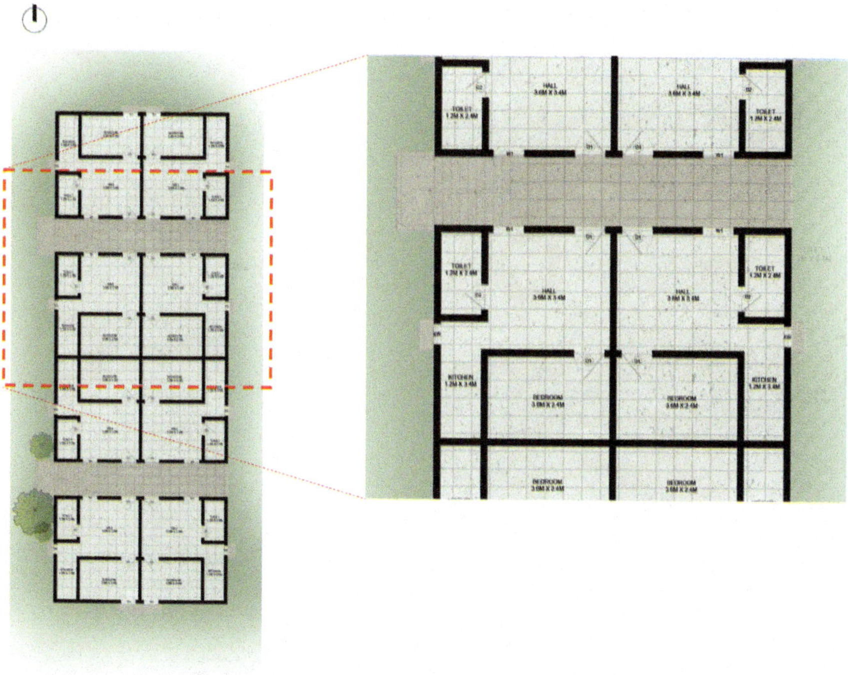

Fig. 10 Typical floor plan of 1 block and blow-up of dwelling unit

In summary, there is a strong demand for well-ventilated dwelling units with improved envelope performance to enhance indoor comfort in both winter and summer, as well as optimal space provision. The Post-Occupancy Evaluation (POE) reveals that while the study area performs within energy use benchmarks, it scores low on liveability, as users report limited appliance usage and restricted usage schedules to manage costs. Instances of excessive power consumption are rare, indicating a promising potential for achieving Net-Zero status with relatively minimal interventions in this case.

For designing a Net Zero Energy community, an energy-efficient built environment is a prerequisite and it can be achieved by predesign decisions at layout level, street level, and building level. The initial design focuses on two parameter sets developed on ENVI-Met: the first set includes main street orientation, building typology, aspect ratio, and surrounding vegetation, offering a filtered selection of optimal options to initiate the design (Fig. 13). The analysis results from the initial parameter set are integrated with a second set (including open space types, building footprint, and surface types) to develop a comprehensive layout for the community.

Preliminary simulations at the layout level indicate that East–West (EW) and Northwest-Southeast (NW–SE) orientations yield better outcomes, though no single building morphology shape clearly outperforms others. For vegetation and Sky View

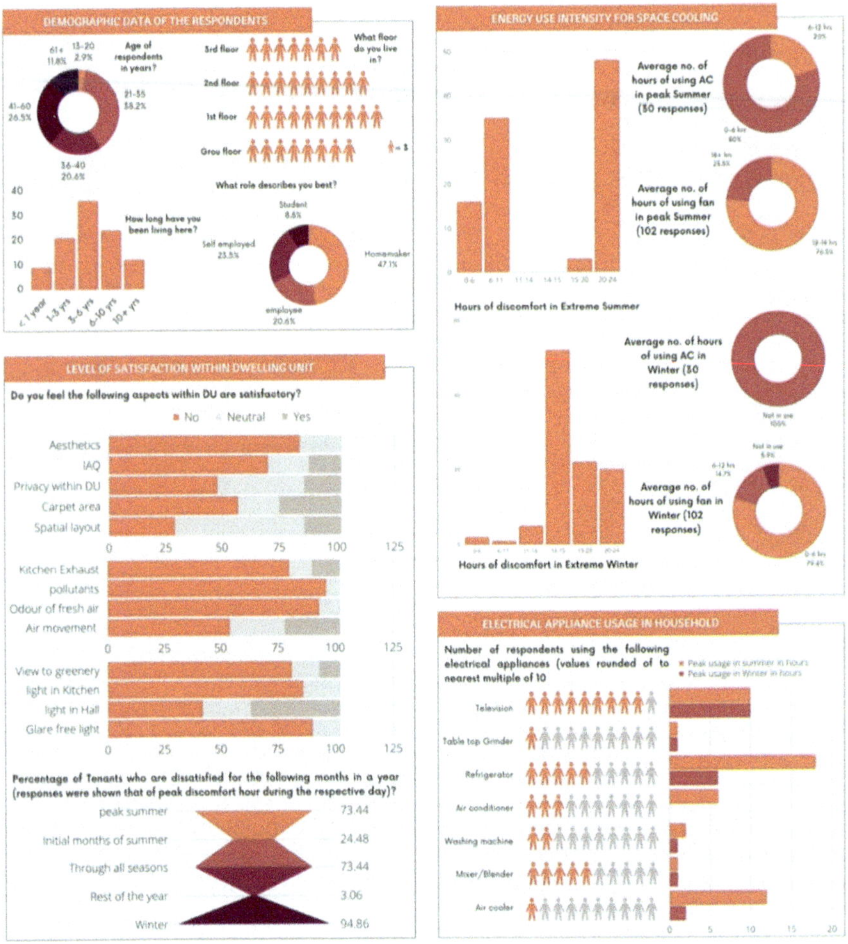

Fig. 11 Survey responses for spatial attributes, thermal comfort, and energy use patterns

Factor (SVF), a stronger impact is observed, while Mean Radiant Temperature (MRT) in winter does not show a consistent correlation with specific parameters.

For Summer—June 21—values at 3.00 pm, as per climate consultant, Wind speed is taken as 3–4 m/s and air temperature is taken as warm (27–38 °C) from South-West.

For Winter—Dec 21—values at 7.00am, as per climate consultant, Wind speed is taken as 1.5–2.4 m/s and air temperature is taken as mild (14–21 °C) from North].

Table 3 highlights the percentage variations of each iteration relative to the benchmark in each group. Selected top-performing iterations are analyzed further, and simulated with the second parameter set.

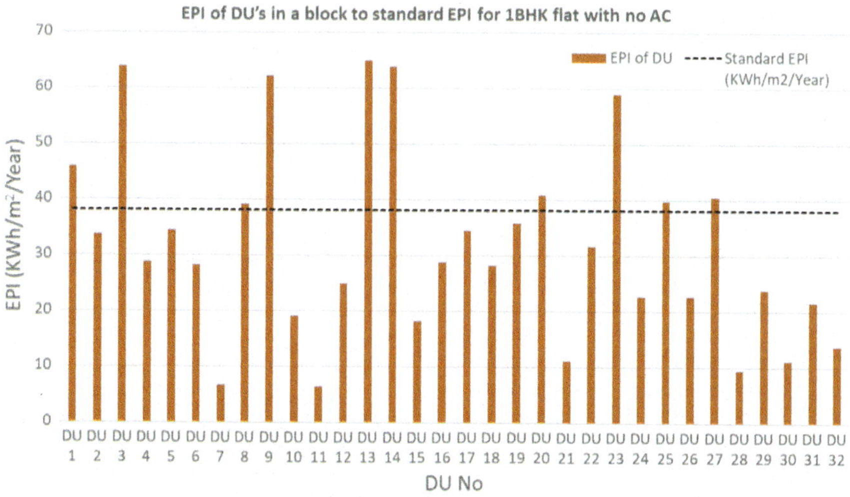

Fig. 12 Base case EPI vs EPI of all dwelling units in one apartment unit

Addressing identified shortcomings at dwelling unit level such as lack of space, lack of proper ventilation and daylight, the prototype dwelling unit is proposed in a way that all the bedrooms and kitchens having windows to exterior and the hall and toilet openings are exposed to the outside. With this as a repetitive dwelling unit, three cases A, B and C as shown in Fig. 14 are developed. Each case varies in open space distribution, cluster arrangement, and balance of soft and hard landscaping.

The PAT comparison of base case and design cases A, B, and C shows that case A improves ambient air within the layout by more than 20%. There is 48% drop in the number of 31.75 °C grids when compared to the base case. The tenants in base case responded that the wind speed in the alleys is very low. In the humid month of August, Fig. 15 also depicts the wind speeds within each cluster. In the design case, the alleys are wide, with a uniform wind speed along the alleys and a greater wind speed at the end of the cluster on the northern side in cases B and C, which induces greater wind speeds within the cluster and aids in naturally ventilating the Dwelling Units. Case B has a higher maximum speed of 6.4 m/s than Cases A and C.

For peak summer conditions observed on June 21 at 2 PM, Case A performs best, maintaining lower potential air temperatures than Cases B and C. Figure 16 reveals that Case A has a lower temperature by up to 0.4 °C across a 150 m × 150 m grid, with reductions of nearly 6 °C at 16 points cumulatively in the layout.

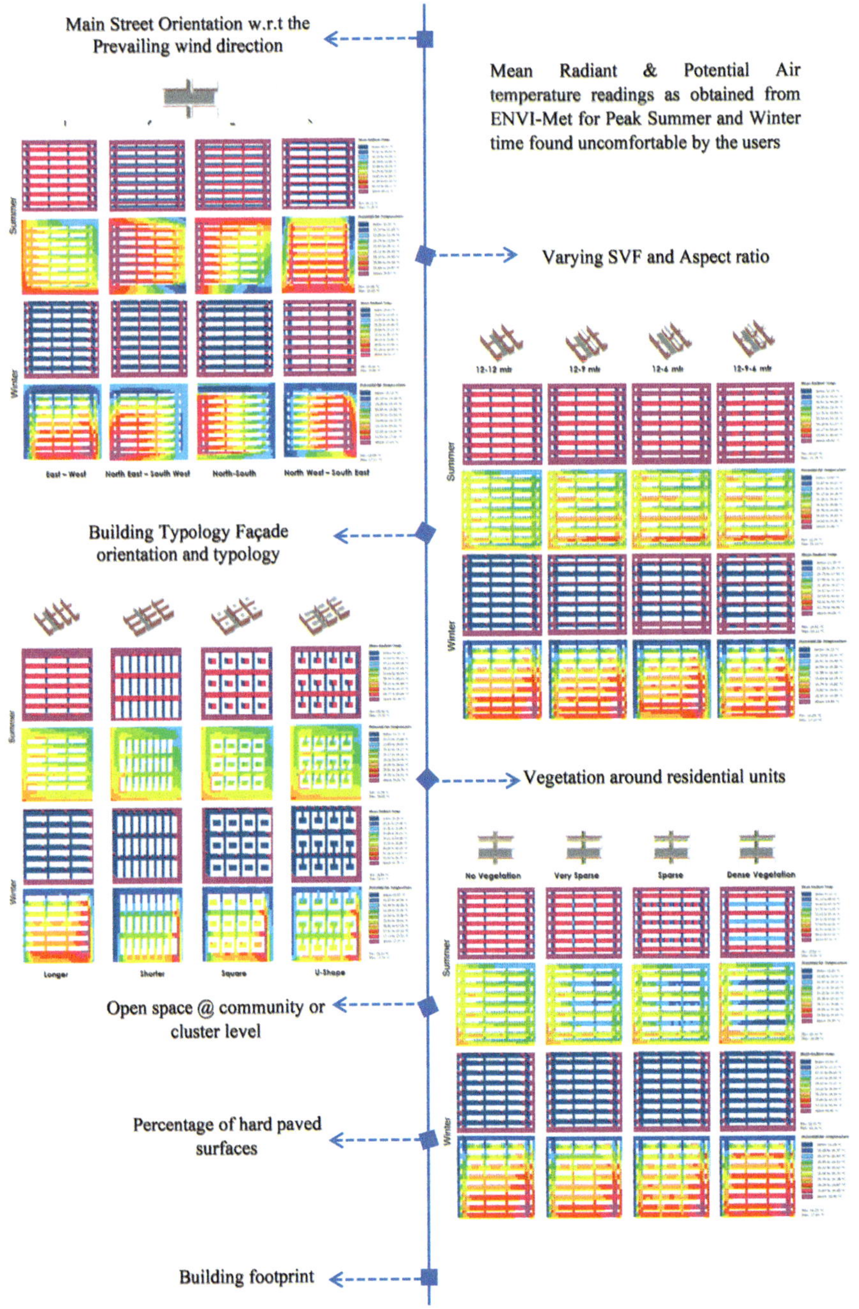

Fig. 13 150 m × 150 m grid with design parameters showing the MRT and PAT in summer and winter, respectively (June 21, 3:00 pm and Dec 21, 7:00 am)

Table 3 Percentage variation among the first set of iterations using ENVI-Met

Measure	Main street orientation Parameter – P1				Building morphology Parameter – P2				Varying SVF with varied ht. Parameter – P3				Level of Vegetation Parameter – P4			
	EW	NE-SW	NS	NW-SE	Longer face	Shorter	Square	U shape	12-12	12-9	12-6	12-9-6	No vegetation	Very sparse	Sparse	Dense
MRT in Summer	BMS_{P1}	9.44%	9.61%	9.37%	BMS_{P2}	8.03%	7.1%	7.1%	BMS_{P3}	0.69%	3.58%	2.54%	BMS_{P4}	0.97%	7.42%	12.98%
MRT in Winter	BMW_{P1}	24.79%	8.26%	123.9%	BMW_{P2}	9.15%	1.61%	1.61%	BMW_{P3}	2.85%	2.85%	3.68%	BMW_{P4}	1.32%	2.47%	7.27%
PAT in Summer	BPS_{P1}	1.57%	1.1%	1.13%	BPS_{P2}	0.14%	0.07%	0.12%	BPS_{P3}	0.55%	0.6%	0.6%	BPS_{P4}	0.12%	0.12%	0.18%
PAT in Winter	BPW_{P1}	0.15%	0.53%	0.79%	BPW_{P2}	0.62%	0.47%	0.20%	BPW_{P3}	0.24%	0.48%	0.32%	BPW_{P4}	0.03%	0.07%	0.52 %

Note: B-Benchmark, M-Mean Radiant Temperature (MRT), P-Potential air Temperature (PAT), S-Summer, W-Winter, P1, P2, P3, P4-Parameter 1,2,3 and 4.

BPS_{P1} stands for Benchmark value of PAT for Parameter 1 in summer

Orange shows that the iteration is bad at performance than its benchmark value in the respective group. Greater the percentage, darker the cell.	Blue infers that the readings of the iteration are considered as a benchmark to compare with the rest of the cases.	Green refers that the iteration is better than its benchmark value in the respective group. Greater the percentage, darker the cell.

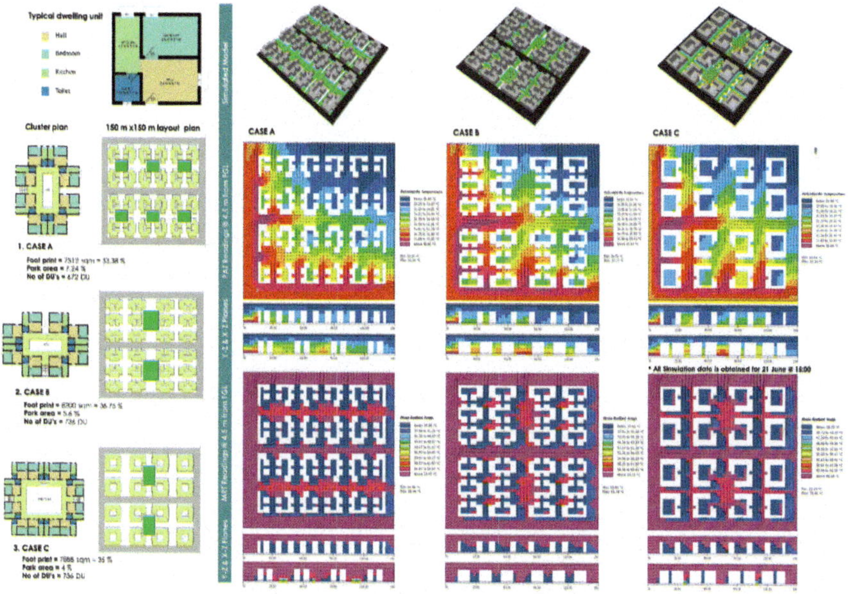

Fig. 14 PAT and MRT values recorded on 21 June at 2:00 PM for iterative cases A, B and C

6 Conclusion and Future Studies

This study demonstrates the importance of integrating climate-responsive design strategies at both the micro (dwelling unit) and macro (layout and street level) scales to enhance indoor comfort and reduce energy consumption. The proposed interventions and the performance comparisons provide a solid foundation for creating a more sustainable and energy-efficient built environment for the economically weaker

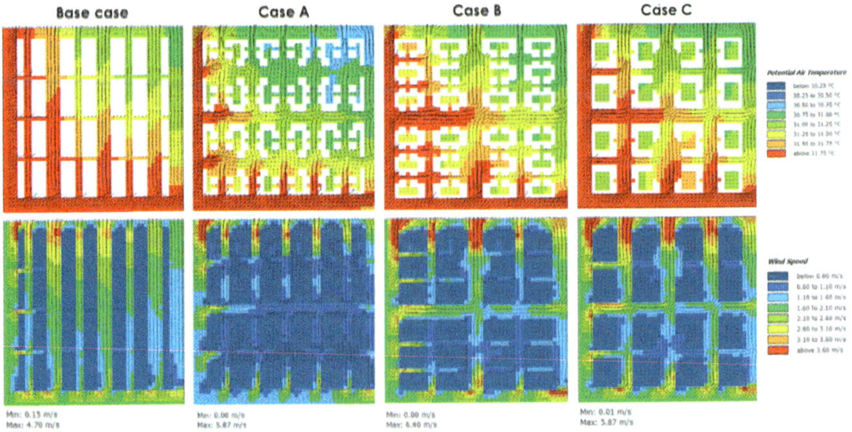

Fig. 15 PAT and wind velocity within 150 m × 150 m layout of base case, case A, B, and C

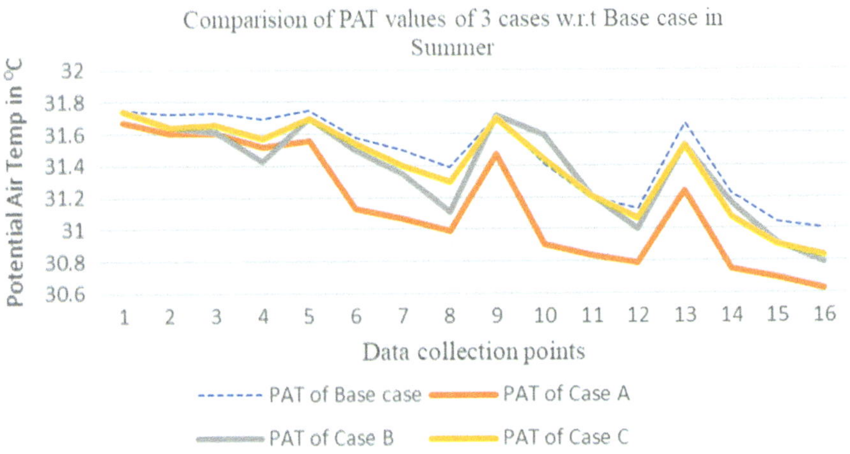

Fig. 16 Comparison of potential air temperature of base case and design cases A, B and C

sections in Visakhapatnam, with potential for replication in similar climates across India.

The study reveals that the built environment aspects of the community are inadequately planned in the base case, but these deficiencies can be effectively addressed through design interventions at the layout, street, and building levels. A modular grid of 150 m × 150 m is designed using seven key parameters, organized into two groups. The first group includes experiments with various orientations, vegetation layouts, and aspect ratios. Simulations in ENVI-met were used to record Mean Radiant Temperature (MRT) and Potential Air Temperature (PAT) readings during typical summer and winter hours. Each parameter was tested against a base case, with performance comparisons for each alternative configuration.

The results indicate that the layout with the main street oriented along the NE–SW direction performs approximately 10% better in winter, while in summer, the N–S street orientation outperforms the base case with an E–W orientation. Among the different building morphologies, the shorter typology proves to be the most beneficial, offering a 0.15% cooler environment compared to the base case, when compared to U-shape, square, and longer façade typologies. In terms of Sky View Factor (SVF) and aspect ratio, the base case, with a building height of 12 m and street width of 12 m, performs better in most instances than the other alternatives. Lastly, the case with the most vegetation shows superior performance over all other vegetation alternatives.

These options are further iterated considering courtyard typology housing and different open space areas at community level namely case A, B, and C and compared with the existing case. In essence, case A reduces Potential air temperature by 48–50% more than the base case, and performing better than the other two design cases. In cases A, B, and C, wind velocity was notably higher compared to the base case. Case A, in particular, demonstrated a 25–30% increase in air velocity, significantly enhancing natural ventilation and improving overall comfort. The typology in base case can be replaced by the built morphology shown in design case A for more comfort at the layout level, improving ambient conditions within the layout and ultimately leading to energy efficiency at the community scale.

This form-based study primarily focused on residential layout and street-level parameters, varying one parameter at a time and recording its impact. However, this approach may have overlooked certain potential layout configurations that could arise from optimizing multiple parameters simultaneously. Future research could expand this approach by exploring "Parametric Urbanism," where all parameters are varied together to achieve an optimized outcome. Additionally, incorporating building comfort parameters, along with considerations for energy supply, water supply, land use diversity, and transit systems, would allow for a more holistic design, ultimately facilitating the creation of a comfortable, energy-efficient, and Net Zero Energy Community.

References

Boarin, P., Besen, P., & Haarhoff, E. (2018). Post-occupancy evaluation of neighborhoods: A review of the literature. In *Working Paper 18-01*. Auckland, New Zealand: National Science Challenge 11: Building Better Homes, Towns and Cities.

Campos, E. P. K. (2020). *Post occupancy evaluation for liveability in urban districts, master's thesis*. University of Stavanger/Technical University of Denmark. https://hdl.handle.net/11250/2680969

Query ID="Q5" Text="" Carlisle, N., Van Geet, O., & Pless, S. (2009). Definition of a 'Zero Net Energy' community (No. NREL/TP-7A2-46065). National Renewable Energy Lab (NREL), Golden, CO (United States).

D'Amico, B., Pomponi, F., & Hart, J. (2021). Global potential for material substitution in building construction: The case of cross laminated timber. *Journal of Cleaner Production, 279*, 123487.

Franks, C. E. I. B. (2020). *NZN: Net zero neighbourhood* (Doctoral dissertation, Worcester Polytechnic Institute).

Kallushi, A., Harris, J., Miller, J., Johnston, M., & Ream, A. (2012). Think bigger: Net-zero communities. In Proceedings of ACEEE (Vol. 11, pp. 115–127).

Mavrigiannaki, A., Gobakis, K., Kolokotsa, D., Kalaitzakis, K., Laura Pisello, A., Piselli, C., Laskari, M., Saliari, M., Assimakopoulos, M.-N., Pignatta, G., Synnefa, A., & Santamouris, M. (2021). Zero energy concept at neighborhood level: A case study analysis. *Solar Energy Advances, 1*, 100002. ISSN 2667-113.

MoHUA, GoI. (2024). PMAY-U Progress Report. Available at-https://pmay-urban.gov.in/uploads/progress-pdfs/660aa6b1d6be5-PMAY-U_Achievement_WEB_01042024.pdf

Integration of Biophilic Design Principles in Development Control Regulations: A Case of Educational and Office Building in India

Lilly Rose Amirtham, Shivanand, and R. Bavya

Abstract Biophilic design connects built environments with nature, fostering a symbiotic relationship that benefits both humans and the planet. In an era marked by rapid urbanization, environmental degradation, and a growing disconnect between humans and the natural world, the integration of biophilic strategies in building design has emerged as a promising solution. Biophilia has gained significant attention as it not only reduces the ecological footprint of buildings but also enhances human well-being and productivity. The incorporation of biophilic elements into architectural designs seeks to create harmonious environments that reconnect occupants with nature and blurs the boundaries between outdoor and indoor areas. However, despite the potential benefits, the practical implementation of biophilic strategies within building design presents a series of complex challenges that need to be addressed. Therefore, this chapter attempts to identify the challenges in integrating biophilic principles in building design, assess the benefits of integration of biophilic design in an educational and office typology, and explore the possibilities of integrating biophilia in development regulations in India. Firstly, a questionnaire survey was conducted to understand the user perceptions and identify the challenges faced by the design professionals in integrating biophilic principles. Then, the benefits of integrating biophilic principles in building design are assessed through case studies in institutional and office buildings. Further, the integration of biophilia in building regulations is explored through assessment of development control codes in various cities in India and abroad. The study revealed that users prefer to have a close connection with nature by integrating medium-density green wall, indoor plants, wooden furnishings and wall colors in the educational environment. The major challenges identified include lack of awareness among clients regarding biophilic design and its associated benefits, and insufficient research data supporting the tangible benefits and effectiveness of biophilic design. The study also found that an increase of 30% green plot ratio enhanced the outdoor thermal comfort by reducing the MRT by 22.94 °C

L. R. Amirtham (✉)
Department of Architecture, School of Planning and Architecture, Vijayawada, India
e-mail: lillyrose@spav.edu.in

Shivanand · R. Bavya
School of Architecture (V-SPARC), Vellore Institute of Technology, Vellore, India

J. Vijayalaxmi (ed.), *Sustainable Building Performance*,
https://doi.org/10.1007/978-981-96-2428-7_7

and PET by 4.25 °C in an office building and the vegetated roofs, reduced the PET and MRT by 17 °C and 43 °C respectively. The findings of the study demonstrated positive benefits of integrating biophilia in building regulations in India as practiced in countries like Singapore, Denmark, etc., thus paving way toward achieving the goals of SDG 11 (Sustainable cities and communities).

Keywords Biophilic design · User perceptions · Outdoor thermal comfort · Energy consumption · Co_2 emissions · Green cover

1 Introduction

Sustainable architecture and Biophilic design have gained significant attention in recent years, as global concerns over environmental degradation and resource depletion escalate. There has been a growing focus on 'Nature' and 'biophilic design' over the past decade, particularly as a response to these environmental issues. Environmental concerns such as climate change, ozone depletion, and loss of biodiversity emerged, and solutions to these issues came to be characterized within the call for sustainable development (Zhong et al., 2022). Biophilia, which connects built environments with nature is viewed as a response to these environmental concerns thereby enhancing the health and productivity of the occupants. Biophilic design bridges the gap between the built environments and nature, fostering a symbiotic relationship that benefits both humans and the planet.

Biophilia is defined as the intrinsic human predisposition to connect with natural processes and systems, which consider the experiences of the natural world as part of the daily lives of humans, whether at home, at work, or during recreation (Kellert & Calabrese, 2015; Wilson, 1992). Biophilic Architecture is based on six biophilic aspects which include environmental characteristics, natural shapes and forms, natural patterns and processes, light and space, place-based interactions, and human–nature relationships (Kellert & Calabrese, 2015). These aspects highlight integrating environmental features such as natural light, vegetation, and water in building design; mimicking natural forms, shapes, and patterns in design; enhancing the design by connecting significant geographical features like mountains, deserts, rivers, ecology, etc., and increasing the affinity through biophilic design configurations (Fig. 1). The connection between human physical, mental well-being, and nature has become increasingly relevant in recent years. The human-oriented planning and design of the built environment, which includes the provision of accessible space, community space, and green space with environmentally-friendly considerations, are fundamentally related to both the physical and spiritual well-being of occupants (Chang & Bae, 2017). Emphasizing such a connection between humans and nature is the concept of Biophilia.

Terrapin Bright Green classified biophilic design into 14 Patterns and documented the health and well-being benefits of these patterns in reducing stress, enhancing cognitive performance, and increasing workplace satisfaction. These 14 patterns

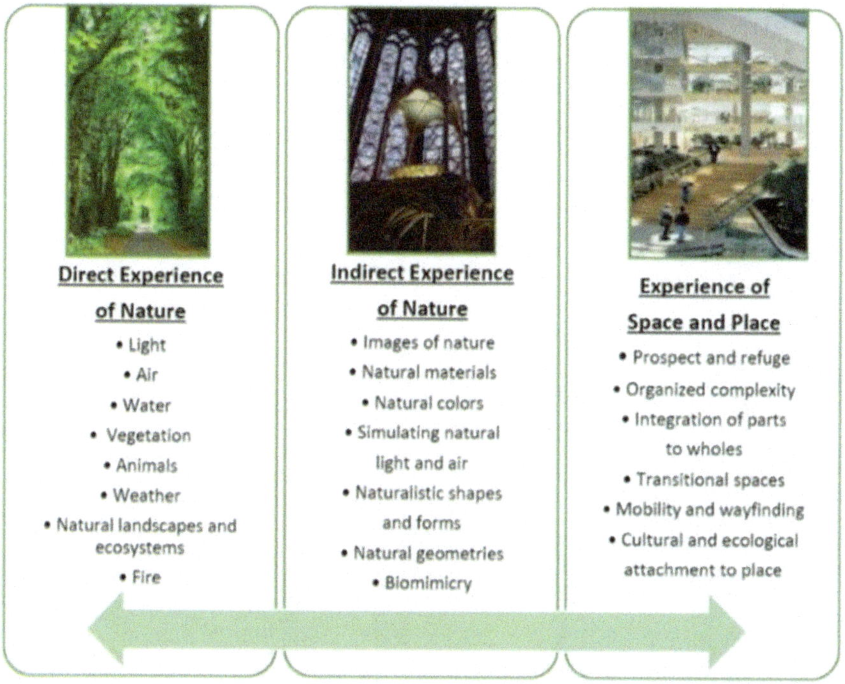

Fig. 1 Experiences and attributes of biophilic design. *Source* Modified from Sevinc Kayihan et al. (2018)

were grouped into three major classifications namely nature in the space, nature analogues, and nature of the space (William Browning et al., 2014). The patterns have been developed through extensive interdisciplinary research and are supported by empirical evidences from research. These 14 patterns have a wide range of applications for both interior and exterior environment, and are meant to be flexible and adaptive, allowing for project-appropriate implementation. (William Browning et al., 2014).

William Browning et al. (2014) analyzed research papers and mapped the availability of rigorous empirical data for supporting the health and well-being impacts of these patterns (Table 1). The study found that three patterns namely well connections with nature, prospect, and refuge had greater impact through peer-reviewed evidences, and two patterns namely connection with natural systems and material connection with nature had no supporting research evidences to explain their relationship with human health and well-being. Many studies have provided evidence of the health and restorative benefits of Biophilia (Frumkin, 2001). Soderlund & Newman (2015) provided evidence of the connection between intrinsic human physiological and psychological reactions to nature, in addition to Biophilia's environmental, economic, and social value in responding to human concerns.

Table 1 Patterns of biophilic design and associated health benefits

14 Patterns		*	Stress reduction	Cognitive performance	Emotion, mood, and preference	Strategy
Nature In the space	Visual connection with nature	***	Lowered blood pressure and heart rate	Improved mental engagement/attentiveness	Positively impacted attitude and overall happiness	
	Non-visual connection with nature	**	Reduced systolic blood pressure and stress hormones	Positively impacted cognitive performance	Perceived improvements in mental health and tranquility	
	Non-rhythmic sensory stimuli	**	Positively impacted heart rate, systolic blood pressure, and sympathetic nervous system activity	Observed and quantified behavioral measures of attention and exploration		
	Thermal and airflow variability	**	Positively impacted comfort, well-being and productivity	Positively impacted concentration	Improved perception of temporal and spatial pleasure (alliesthesia)	

(continued)

Table 1 (continued)

14 Patterns		*	Stress reduction	Cognitive performance	Emotion, mood, and preference	Strategy
	Presence of water	**	Reduced stress, increased feelings of tranquility, lower heart rate and blood pressure	Improved concentration and memory restoration Enhanced perception and psychological responsiveness	Observed preferences and positive emotional responses	
	Dynamic and diffuse light	**	Positively impacted circadian system functioning Increased visual comfort			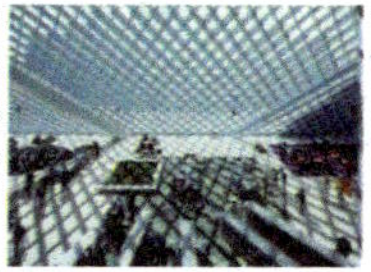
	Connection with natural systems				Enhanced positive health responses; Shifted perception of environment	
Natural analogues	Biomorphic forms and patterns	*			Observed view preference	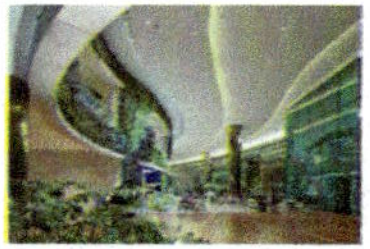

(continued)

Table 1 (continued)

14 Patterns		*	Stress reduction	Cognitive performance	Emotion, mood, and preference	Strategy
	Material connection with nature			Decreased diastolic blood pressure Improved creative performance	Improved comfort	
	Complexity and order	**	Positively impacted perceptual and physiological stress responses		Observed view preference	
Nature of the space	Prospect	***	Reduced stress	Reduced boredom, irritation, fatigue	Improved comfort and perceived safety	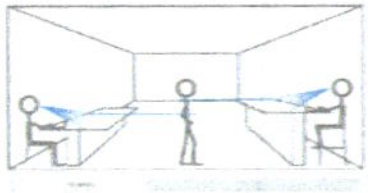
	Refuge	***		Improved concentration, attention, and perception of safety		

(continued)

Table 1 (continued)

14 Patterns		*	Stress reduction	Cognitive performance	Emotion, mood, and preference	Strategy
	Mystery	**			Induced strong pleasure response	
	Risk/Peril	*			Resulted in strong dopamine or pleasure responses	

Source Modified from William Browning et al. (2014)
Note (*)—indicates the availability of quantity and quality peer-reviewed evidences

1.1 Biophilia and SDGs

The concept of sustainable development was developed by the United Nations (UN) Brundtland Report (1987) and was further elaborated through Agenda 21 (1992) and the 17 Sustainable Development Goals (2015). Biophilic patterns serve as a sustainable means to achieve several SDGs such as SDG 3(Good health and well-being); SDG 7(Affordable and Clean energy); SDG 9 (Industry Innovation and Infrastructure); SDG 11 (Sustainable Communities and Cities); SDG 12 (Responsible Consumption and Production); SDG 13 (Climate Action); and SDG 15 (Life on Land). The building sector plays an essential role in sustainable development and is responsible for nearly 40% of energy consumption and energy-related carbon dioxide emissions (Ibrahim & Al-Chaderchi, 2022). Buildings also have a significant impact on human health and well-being, as we spend approximately 90% of our time indoors. In addition, the outbreak of the COVID-19 pandemic and stay-at-home mandates increased people's attention to mental and psychological well-being and highlighted the role of the building design on people's health and well-being. The issues with the COVID-19 pandemic (such as contact with nature, improved indoor air quality, and comfort) emphasized the opportunity to link building design with nature instead of focusing only on improving energy efficiency in buildings (Nitu et al., 2022).

Biophilia reduces the ecological footprint of buildings while enhancing human health and productivity. Integration of biophilic elements into architectural designs helps in re-establishing the harmonious connection of the occupants with nature that existed before urbanization thereby blurring the boundaries between outdoor and indoor areas. However, despite the potential benefits, the practical implementation of biophilic strategies within building design presents a series of complex challenges that need to be addressed. Therefore, this chapter attempts to identify the challenges in integrating biophilic principles in building design, assess the benefits of integration of biophilic design in an educational and office typology, and explore the possibilities of integrating biophilia in development regulations in India.

2 Implementation of Biophilic Strategies

The influence of built environment on occupant's well-being and productivity has been well researched in various typologies such as offices, hospitals, housing communities, educational institutions, etc. Aristizabal et al. (2021) identified the potential benefits of incorporating nature-based design elements in office environments, such as stress reduction, improved cognitive performance, and increased workplace satisfaction. Also, adding natural features to retail and service settings boosted the customer engagement and lower anxiety (Castro et al., 2022) and integrating biophilic design in healthcare facilities by incorporating indoor plants enhanced occupant's health and well-being (Moslehian et al., 2023). Colleges and universities can take advantage of nature-inspired themes to positively support students thereby creating environments that reduce stress, enhance creativity and cognitive development (DeLauer et al., 2022; Martin et al., 2020; Yusufov et al., 2019). Abdelaal (2019) found that the principles of biophilia have a greater potential in fostering innovation in higher education. Although there are several psychological benefits associated with integrating biophilic design, barriers such as high costs, tangible benefits, knowledge limitation, and maintenance burden also exist (Sadick et al., 2023). Also, the impacts of biophilic design on occupants' health and the challenges for implementing biophilic strategies in Indian scenario are less explored. Hence, this chapter attempts to present the findings of user perception survey of university students in a Green Rated Educational Campus and survey conducted among the design professionals to identify the challenges encountered in incorporating biophilic design in India.

2.1 User Perception Survey in an Educational Institution

Biophilic patterns help in mitigating the negative effects on the student's quality of life. Advantages of integrating biophilic principles into campus design stimulates and sustains the creative and imaginative capacities of users (Abdelaal & Soebarto, 2019). Two Green Rated Educational Institutions namely Vellore Institute of Technology (VIT) and Koneru Lakshmaiah University (KLU) were selected for the study. A questionnaire survey with 22 questions that include information on general data, nature connectedness, location and accessibility, thermal comfort, visual comfort, auditory comfort, Indoor Air Quality (IAQ), interior design, visual connectivity to nature (direct and indirect), and preferences toward biophilia. A total of 88 students (43 students from VIT and 45 students from KLU) were surveyed and the survey results are presented in Table 2. The survey results revealed that students preferred to have small waterbody, good natural views from the windows, calm zones, aromatic spaces, functional living walls and greenery, biophilic form and façade, colorful interiors, easy accessibility, green outdoor spaces for leisure activities. The findings of the study reinforce that biophilic implementations should not only be considered for classrooms but should include other learning environments as they enhance the human-nature relationships and physical activities impacting quality of life on campus. Transition spaces and areas in between classrooms and buildings aid in enriching place-based interactions and experience. This is similar to the findings of Peters and D'Penna (2020) which insisted that designers should incorporate outdoor green space between buildings to facilitate social interaction and encourage green views from the building.

The responses of the questionnaire survey revealed that seven elements were preferred by the students in the educational buildings. They are green walls (98%), shaded outdoor spaces for leisure activities (89%), water area (75%), biophilic form and facade designs (72%), colorful interior spaces (65%), aromatic spaces (56%), and sounds that mimic nature (44%). Based on the above findings, the existing design

Table 2 Findings of user perception survey with regard to biophilia

Biophilia aspects		VIT (IGBC platinum certified)	KLU (GRIHA 5 star rated)
Biophilic infrastructure			
1	*Enlarging water area (presence of waterbody)*	**75%** of the participants opted for a small waterbody	**76%** of the participants opted for a small waterbody
Sensorial design			
2	Visual connection with nature through window views	**71%** of the participants were satisfactory	**58%** of the participants were satisfactory
3	Natural sound design	**44.2%** opted for sounds that mimic nature	**22.2%** opted for sounds that mimic nature
4	Acoustic comfort	**97%** of the students were dissatisfied with the noise levels inside the studio	**80%** of the students were satisfied with the noise levels inside the studio
5	*Aromatic plant design*	**55.8%** prefer aromatic spaces	**44.4%** prefer aromatic spaces that mimic nature
6	Thermal comfort	Majority of the respondents are **comfortable** with the thermal comfort provided through passive strategies	Majority of the respondents are **comfortable** with active and passive strategies
7	Visual comfort	**84%** were satisfied with the daylight conditions however 16% reported that it is not uniformly spread across the studio due to its large span	**90% of the respondents are comfortable** with the lighting conditions
8	*Provision of indoor plants and external green walls (vertical garden)*	**98%** prefer functional living walls and greenery	**91%** prefer functional living walls and greenery
Biophilic setting and performance			
9	*Biophilic form and facade design*	**72%** prefer biophilic form and facade designs	**36%** prefer biophilic form and facade designs
10	Interior surface finishes and furnishings	Students prefer **colorful interior surface finishes** that enhances creativity to monochrome finish	Students are satisfied with the interiors
Connectivity			
11	Accessibility	Building is situated 1 km from the entrance. Not easily accessible	Students reported ease of access
Green space place making			

(continued)

Table 2 (continued)

Biophilia aspects		VIT (IGBC platinum certified)	KLU (GRIHA 5 star rated)
12	*Preferences for shaded outdoor spaces (leisure activities)*	**88.4%** of the participants prefer to have outdoor leisure activities	**62.2%** of the participants prefer to have outdoor leisure activities
13	*Preferences of biophilic elements*	**54%** preferred biophilic form and facade, natural sound and aroma, outdoor green spaces for leisure activities	

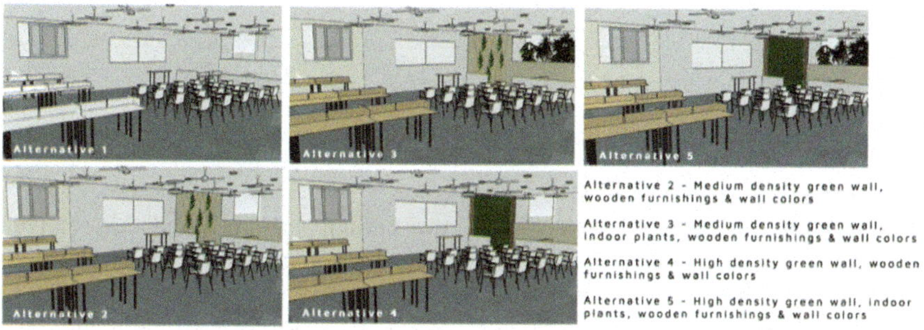

Fig. 2 Alternative design solutions

studio (base case—alternative 1) was enhanced by integrating biophilic design solutions such as good natural views, provision of waterbody, indoor plants, colorful interiors, natural materials for finishes and furnishings, and usage of plants as a noise buffer for improve acoustic comfort, and alternative design options were developed (alternatives 2—5) as shown in Fig. 2. To analyze the satisfaction level of the previously identified biophilic attributes, the users of an architectural design studio in VIT (sample size of 51) were asked to assess these options through a questionnaire. 54.9% of the participants were male and 45.1% were female.

IBM SPSS Statistics tool has been used in the study to analyze the user perception survey response. The responses were tested for significant correlations between the selected attributes and the highest mean of the proposed redesign. Spearman correlation was used to explore the difference in average and relationship between the proposed design and biophilic attributes to achieve higher user satisfaction. Insignificant correlations were neglected from the results. In the correlation analysis section, p was set to 0.05. Five variables namely visual comfort (VC), natural ventilation (NV), indoor plants (IP), acoustic comfort (AC), and interior design for creativity (ID) were assessed in the study. A measurement consistency coefficient of 0.9 was achieved in Cronbach's Alpha, which showed that the ratings were considered reliable with very good consistency. Hence, the data was used to draw conclusions using further statistical analysis.

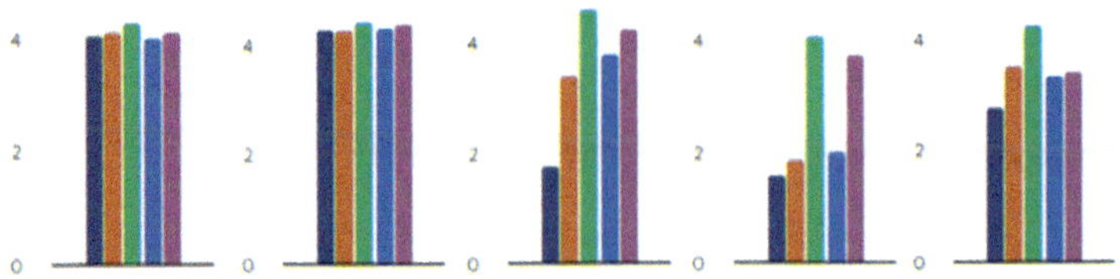

	VC	NV	IP	AC	ID
A1	4.08	4.31	1.78	1.59	2.80
A2	4.14	4.31	3.45	1.88	3.53
A3	4.31	4.47	4.67	4.12	4.27
A4	4.04	4.35	3.82	2.02	3.35
A5	4.12	4.43	4.29	3.75	3.45

Fig. 3 Comparison of means of the five biophilic attributes of alternatives

The survey responses revealed that the participants preferred Alternative 3 (Medium-density green wall, indoor plants, wooden furnishings, and wall colors) with the highest mean of $M = 4.41 \pm 0.8$ (Fig. 3). Alternative 3 was rated the highest in all five biophilic attributes. The participants majorly preferred medium-density living wall to high-density living wall. The VC and NV were very satisfactory in the existing design (A1) as it is a green-rated building (Fig. 4). Comparatively higher correlations were seen between IP and ID. The IP also had a moderate correlation on the user's satisfaction with acoustic comfort. Meanwhile, the existing design for NV has a negative correlation with acoustic comfort. Although the user perception study revealed that integrating biophilic attributes in building increased the satisfactory level of the users, professionals find it challenging to integrate biophilic strategies in design.

2.2 *Professional Challenges in Implementing Biophilic Strategies*

A comprehensive survey was conducted among design professionals to identify the challenges in implementing biophilic design. 70 professionals with more than 2 years of working experience participated in this survey. The challenges identified through the survey have been categorized into major and minor challenges faced in the profession (Fig. 5). The primary influencing factors include clients, designers, innovation, and socio-political influencers. Among the major challenges, a noteworthy concern is the lack of awareness among clients regarding biophilic design and its associated benefits. Additionally, there is a dearth of research or data supporting the effectiveness of biophilic design, leaving professionals without substantive evidence of the environmental and personal advantages it offers. Budget constraints, along with

			VC	NV	IP	AC	ID
Spearman's rho	VC	Correlation Coefficient	1.000	.516**	.021	-.024	-.073
		Sig. (2-tailed)	.	<.001	.882	.868	.610
		N	51	51	51	51	51
	NV	Correlation Coefficient	.516**	1.000	-.143	-.185	.118
		Sig. (2-tailed)	<.001	.	.317	.194	.410
		N	51	51	51	51	51
	IP	Correlation Coefficient	.021	-.143	1.000	.499**	.623**
		Sig. (2-tailed)	.882	.317	.	<.001	<.001
		N	51	51	51	51	51
	AC	Correlation Coefficient	-.024	-.185	.499**	1.000	.295*
		Sig. (2-tailed)	.868	.194	<.001	.	.035
		N	51	51	51	51	51
	ID	Correlation Coefficient	-.073	.118	.623**	.295*	1.000
		Sig. (2-tailed)	.610	.410	<.001	.035	.
		N	51	51	51	51	51

**. Correlation is significant at the 0.01 level (2-tailed).

*. Correlation is significant at the 0.05 level (2-tailed).

Fig. 4 Spearman's correlations of five biophilic attributes

maintenance and upkeep concerns, emerge as other significant challenges faced by both clients and professionals during the implementation of biophilic designs. In addition to these major challenges, various minor challenges have been identified. These include the limited availability of suitable biophilic materials, challenges related to compliance with building codes and regulations, a limited understanding of construction techniques for biophilic design, mindset-related obstacles, influence from builders, concerns about liability for design risk, employment circumstances, and the need for adapting to changes in living styles. Moreover, the survey revealed that limited availability of suitable biophilic materials poses a hurdle, as sourcing materials that align with the principles of biophilic design can be challenging. Compliance with building codes and regulations emerged as a significant concern, highlighting the need for a harmonious integration of biophilic elements within the existing regulatory frameworks. Professionals also expressed a limited knowledge about construction techniques specific to biophilic design, pointing toward the importance of further education and training in this emerging field. The survey brought to light the impact of mindset-related challenges, builder influence, and concerns about liability for design risk, underscoring the need for a shift in perspectives and enhanced collaboration within the industry. Moreover, employment circumstances and the evolving nature of living styles were identified as minor challenges, emphasizing the dynamic nature of the field and the need for professionals to adapt to changing preferences and lifestyles. This comprehensive exploration of challenges aims to inform strategies and interventions that can facilitate the successful integration of biophilic design principles in the building industry.

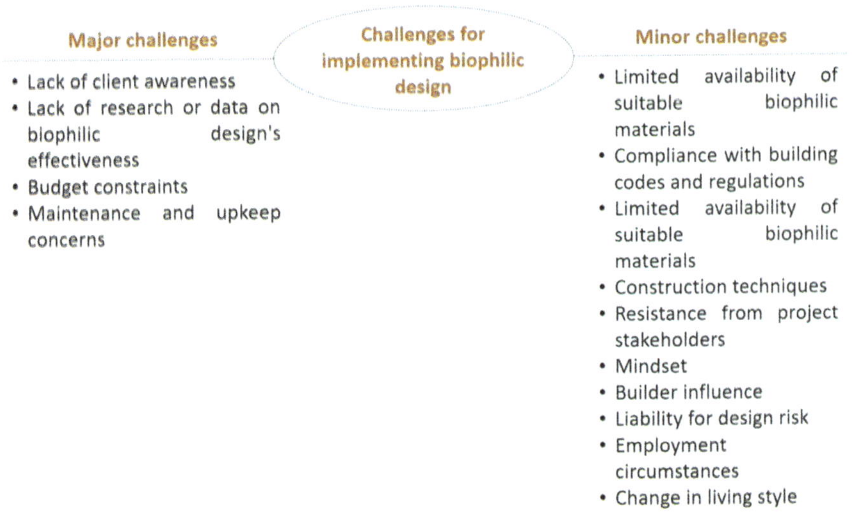

Fig. 5 Challenges for implementing biophilic design

3 Benefits of Integration of Biophilic Design

Integration of biophilic strategies in the built environment enhances the quality of life in cities. The user perception analysis (2.1) revealed that biophilic strategies enhance human–nature relationships and enrich place-based interactions. In spite of these benefits, lack of research data on tangible benefits of biophilic strategies hinders the adoption of biophilia in built environments. In order to quantify the benefits of biophilia strategies, the authors analyzed the impacts of green space on outdoor environment. An office building in Noida was selected for the analysis. Field measurements were collected for 48 h at one-hour frequency on 9th and 10th February 2024. Envi-met software was used to analyze the impact of built environment on outdoor thermal comfort. The existing configuration of the office building was modeled and simulated in envi-met which is considered as the base case for the analysis. The simulated data was validated using the field measurements (Fig. 6). The regression analysis revealed a significant correlation of $R^2 = 0.77$. Hence, Envi-met was used to analyze alternative scenarios by integrating 30% green plot ratio at the site level. The simulations were run for a typical summer day (21.06.2024). The impact of GnPR on air temperatures, relative humidity, mean radiant temperatures (MRT), and physiological equivalent temperatures (PET) were analyzed to quantify the benefits of biophilia.

Figure 7 shows the air temperature, MRT, and PET variations between the base case and the modified case by integrating biophilic strategy. Integration of 30% green cover reduced the air temperatures by 0.78 °C, MRT by 22.94 °C, PET by 4.25 °C, and increased the humidity by 5.61%. This significant reduction in MRT and PET shows the potential to mitigate the urban heat island and improve overall thermal

SUMMARY OUTPUT

Regression Statistics	
Multiple R	0.878014
R Square	0.770909
Adjusted R Square	0.760949
Standard Error	0.271682
Observations	25

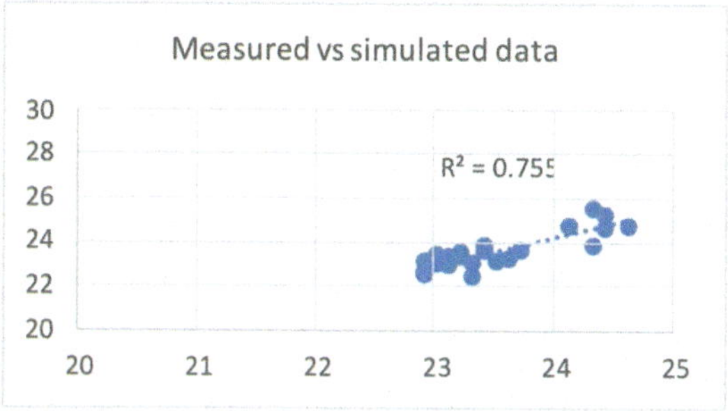

Fig. 6 Validation of envi-met simulated data

comfort. Further, a green roof was compared with exposed concrete roof. The study revealed minimum variations in air temperatures. However, a significant reduction was observed in the PET and MRT with a maximum difference of 17 °C and 43 °C, respectively. PET measures the thermal sensation experienced by individuals. A reduction of 17 °C in PET indicates that the shaded green roofs provide a more comfortable environment in terms of perceived temperature compared to exposed roof areas. This suggests that people would feel cooler and more comfortable in the shaded green roof areas despite similar air temperatures. MRT is a crucial factor in human thermal comfort, as it represents the average temperature of all surfaces surrounding an individual. A reduction of 43 °C in MRT indicates that the surfaces in shaded green roof areas absorb less heat and consequently radiate less heat toward occupants compared to exposed roof areas. This reduction in MRT contributes significantly to the perceived comfort level, as individuals are less affected by radiant heat from their surroundings. PET and MRT highlight the importance of green infrastructure in mitigating urban heat island effects and improving outdoor comfort in urban environments.

4 Integration of Biophilic Design in Development Regulations

Biophilic design principles have been included in various Green Building Rating Tools (GBRTs), to improve the health and well-being of the occupants. Jiang et al (2020) compared the building-level biophilic strategies with the credits of six different international Green Building Rating Tools (GBRTs) namely Leadership in Energy and Environmental Design (LEED) V4, Building Research Establishment Environmental Assessment Method International New Construction (BREEAM International NC) 2016, Green Mark Non-residential Building (GM NRB) 2015, Green Building Label (GBL) 2014, Well Building Standard (WBS) V2, and Living Building Challenge (LBC) V3.1, to identify their compatibility and divergence and developed a framework for revising the GBRTs by including the ignored biophilic

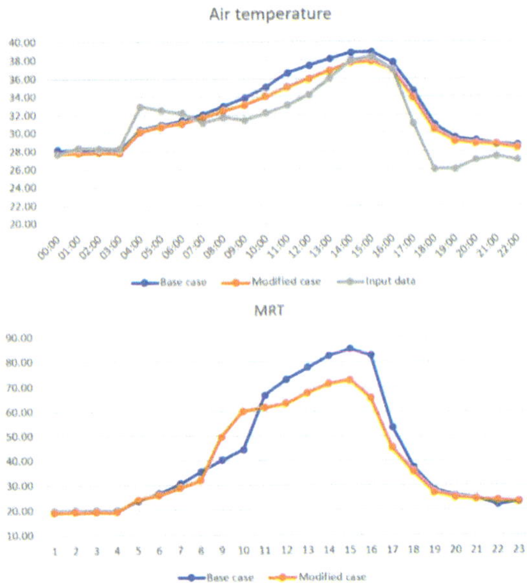

	Reduction			Increase
Time	Potential Air Temperature (°C)	Mean Radiant Temp. (°C)	PET* (°C)	Relative Humidity (%)
00:00	0.24	0.61	0.28	4.90
01:00	0.14	0.66	0.35	3.99
02:00	0.09	0.68	0.39	3.56
03:00	0.05	0.71	0.42	3.28
04:00	0.07	0.28	0.01	1.54
05:00	0.13	1.41	0.57	1.96
06:00	0.10	3.67	1.63	2.37
07:00	0.07	6.46	2.97	2.76
08:00	0.07	8.98	4.25	3.22
09:00	0.26	10.72		4.31
10:00	0.35	22.94	3.84	4.64
11:00	0.73	18.88	1.43	5.61
12:00	0.54	18.83	1.51	4.35
13:00	0.37	21.09	1.78	3.60
14:00	0.26	20.49	1.78	2.69
15:00	0.08	17.86	1.53	2.34
16:00	0.07	1.39	0.23	2.24
17:00	0.07	0.14	0.18	3.24
18:00	0.20	1.17	0.70	1.70
19:00	0.24	1.29	0.77	1.18
20:00	0.21	1.25	0.74	1.27
21:00	0.17	1.30	0.75	1.01
22:00	0.14	1.42	0.80	1.32
23:00	0.12	0.61	0.09	1.01

Fig. 7 Air temperatures, MRT, PET, and RH variations

strategies to improve the health and well-being of the occupants. Amirtham and Bavya (2023) analyzed the integration of biophilic features in three GBRTs namely LEED, Indian Green Building Council (IGBC), and Green Rating for Integrated Habitat Assessment (GRIHA), and found that sustainable site planning and indoor environmental quality (IEQ) of GBRTs are closely associated to six categories of biophilic framework namely biophilic infrastructure, sensorial design, biophilic setting and performance, connectivity, work-life-play relationship and green space place making. The study found that LEED and IGBC have integrated 83% of the biophilic strategies with respect to biophilic infrastructure. Sensorial design, biophilic setting and performance, and green space place making has been ignored in all three GBRTS. The findings revealed that LEED, IGBC, and GRIHA have integrated 58%, 53%, and 36% of biophilic strategies, respectively. Although biophilia has been integrated in GBRTs, its impact on built environment is minimal as they are optional. Integration of biophilic strategies in development regulations is important as it is mandatory for the buildings to adhere to the development control regulations.

Few foreign countries such as Singapore, Denmark, etc., have integrated Biophilic design into building development regulations. However, the building development regulations in India, currently do not have any provisions for biophilic design. Therefore, this study compared the building development regulations of Singapore with major Indian metropolitan cities like Noida and Bangalore specifically for commercial buildings with an emphasis on the integration of biophilic principles. This comparative analysis seeks to uncover points of similarity and difference, offering valuable insights into the possible integration of biophilic strategies in the Indian context. Table 3 shows the integration of biophilia strategies in the guidelines of Singapore. Table 4 shows the guideline criteria of Indian cities for integration of biophilia strategies.

The comparison of Tables 3 and 4 provides a basis for adoption of successful practices of biophilia integration in Indian building development regulations. Figure 8 shows the possible opportunities for integration of biophilic features in building bye-laws.

Based on the above, the following recommendations are suggested for integration of biophilic elements in building development regulations:

- Mandatory green belt in setbacks and in road widening.
- Mandatory GnPR (Green Plot Ratio) for the commercial development with bonus (additional) FAR for fulfilling the 100% GnPR requirement.
- Minimum green belt in setback area with specified LRA to be maintained.
- Cycling plan can be compulsory for project area above 20,000 m^2.
- Communal spaces or interactive spaces restricted to 30–40% of site.
- Water bodies strategies for protecting existing water body or integration of small water elements in design.

5 Conclusion

Integration of small waterbody, good natural views, calm zones, aromatic spaces, functional living walls and greenery, biophilic form and façade, colorful interiors, easy accessibility, green outdoor spaces for leisure activities enhance the user perception significantly in an educational campus. The findings of the study reinforce that biophilic implementations should enhance human-nature relationships and place-based interactions and experience. Integrating a Green Plot Ratio (GnPR) of 30% into building regulations can transform urban landscapes by combining intensive and extensive green roofing techniques. This initiative extends beyond the evident advantages of reducing outdoor temperatures and thermal comfort but also helps in curbing CO_2 emissions, combating urban heat islands, and managing stormwater. It also yields invaluable intangible benefits, such as enriched biodiversity and enhanced mental health and well-being. The integration of green roofs at the building scale transforms the concrete jungle of urban areas into vibrant and sustainable living environments. These green spaces act as natural air purifiers, absorbing pollutants and particulate matter, thereby improving air quality and reducing the risk of respiratory illnesses among inhabitants. Additionally, the presence of vegetation helps to dampen noise pollution, creating quieter and more peaceful urban settings. While there may be initial concerns about the potential rise in maintenance costs associated with implementing green roof systems, these investments are overshadowed by the multitude of environmental and personal benefits they bring. Therefore, it is imperative to integrate biophilia strategies into development control regulations to create sustainable and resilient communities paving way toward achieving the goals of SDG 11.

Table 3 Biophilia integration in Singapore guidelines

S. No.	Parameter	Guideline
1	Bonus gross floor area (GFA)	Bonus GFA schemes applicability: • Central business district incentive scheme • Strategic development related incentive scheme • Facilities for sports and community scheme • Landscaped rooftops as open recreational area (ORA) • Transformation scheme on built environment • Public spaces with private ownership having ORAs
2	Road buffer	Type 1: 15 m (5 m green)—expressway Type 2: 7.5 m (3 m green) major arterial Type 3: 5 m (3 m green) minor arterial Type 4 and 5–5 m (3 m green) primary access
3	Building setback from boundary	As per the plot size required building setback is 6 and 4.5 m (minimum)
4	Building height and additional height allowed	No storey-height control excluding sites that has (a) conservation or urban design requirements (b) technical height controls Floor to floor height maximum 5 m The height of the building must adhere to technical height restrictions set by other regulatory bodies
5	Walking and cycling plan	$\geq$ 20,000sqm GFA
6	Boundary walls/Fences	Chain-link fences no higher than 1.8 m, with closely spaced plants and other vegetation acting as screening. Within the development, a planting strip with a minimum width of 0.5 m must be included
7	Sky terrace area as % of floor plate	More than or equal to 60% of the floor plate is taken up by the sky terrace area inside the 45-degree line
8	Perimeter openness	The sky terrace must have a minimum of 40% of its perimeter left open and unenclosed
9	Landscape replacement area (landscaped areas provided on the upper levels of the development.)	100% in developments in strategic areas 70% in developments in strategic areas outside central area with height control less than or equal to 80 m
9	Minimum softscape requirement as a % of site area	40%
10	Minimum Green plot ratio (GnPR) requirement (Total leaf area/site area)	4
11	Communal pavilions	A pavilion at ground level is restricted to 50 m^2. Its coverage will be applied to the highest (40%) control of overall site coverage. Maximum 500 m^2 at the top of the roof, or 50% of the roof, whichever is less

Source ura.gov.sg

Table 4 Guidelines of Indian cities for integration of biophilia strategies

Criteria	Bangalore	Noida	Included/scope for inclusion
Landscape	In locations where the site size is more than 200 m^2, a minimum of two trees must be planted	One tree per 100 m^2. And 25% minimum open space to be kept for landscaping	
Lighting and ventilation	Opening size should be 1/5 of the floor area One or two windows can be provided. The size of opening shall be increased by 25% in case of kitchen. For washroom the minimum window and ventilator size shall be 0.9 m^2 with one side opening not less than 0.3 m^2	Light and ventilation shall be ensured (National Building Code, 2005)	Direct nature integrated with development regulation
Ground coverage	65, 60, 55% as per site area	30% of site area	
FAR	1, 1.25, 1.5, 1.75 as per ground coverage	2.00 as per ground coverage	Scope where we can integrate nature in development regulation
Setback	According to building height Up to 9.5 to 50 mt—4.5—16 m For smaller heights—1.5–4.5 m	Varies with the land size up to 15 m^2 plot size minimum front setback 3.0 m and for more than 40,000 m^2 land size 16 m	
Road width	9, 12, 15, 18 m, as per the type of development (intense, moderately, sparsely developed area)	–	
Building height	The setbacks, coverage, and FAR limitations shall determine the building's height	No height restriction; the FAR's limitations will control the building's height	

Source Noida and Bangalore building bye-laws

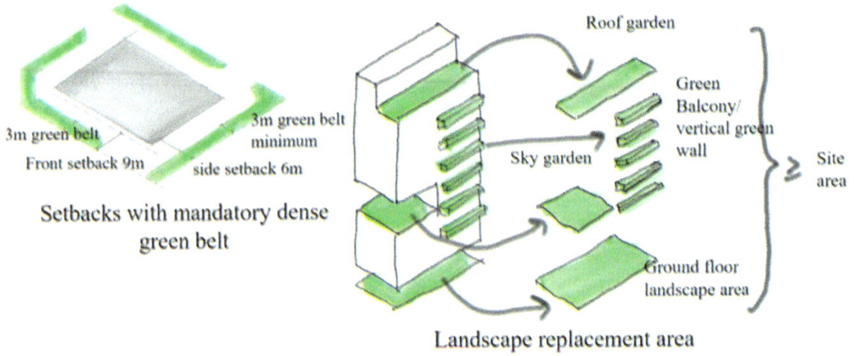

Fig. 8 Recommendation for including green in building bye-laws

References

Abdelaal, M. S., & Soebarto, V. (2019). Biophilia and Salutogenesis as restorative design approaches in healthcare architecture. *Architectural Science Review, 62*(3), 195–205.

Amirtham, L. R., & Bavya, R. (2023). Assessment of integration of biophilic features in green building rating systems: A comparison of Indian and Oman Rating Systems, Green Energy International Conference, 15–16 February 2023, organized by University Technology and Applied Sciences, Salalah.

Aristizabal, S., Byun, K., Porter, P., Clements, N., Campanella, C., Li, L., Mullan, A., Ly, S., Senerat, A., Nenadic, I. Z., Browning, W. D., Loftness, V., & Bauer, B. (2021, October). Biophilic office design: Exploring the impact of a multisensory approach on human well-being. *Journal of Environmental Psychology, 77*, 101682. https://doi.org/10.1016/j.jenvp.2021.101682

Castro, I. A., Honea, H., Cornelis, E., & Majmundar, A. (2022). The friluftsliv response: Connection, drive, and contentment reactions to biophilic design in consumer environments. *International Journal of Research in Marketing, 39*(2), 364–379.

Chang, P. J., & Bae, S. Y. (2017). Positive emotional effects of leisure in green spaces in alleviating work–family spillover in working mothers. *International Journal of Environmental Research and Public Health, 14*(7). https://doi.org/10.3390/ijerph14070757

DeLauer, V., McGill-O'Rourke, A., Hayes, T., Haluch, A., Gordon, C., Crane, J., Kossakowski, D., Dillon, C., Thibeault, N., & Schofield, D. (2022). The impact of natural environments and biophilic design as supportive and nurturing spaces on a residential college campus. *Cogent Social Sciences, 8*(1), 2000570.

Frumkin, H. (2001). Human health and the natural environment. *20*(3).

https://www.ura.gov.sg/-/media/Corporate/Guidelines/Development-control/Commercial/Summary-Commercial.pdf

Ibrahim, I. A., & Al-Chaderchi, B. M. (2022). Contribution of the biophilic design approach to the un sustainable development goals. *WIT Transactions on the Built Environment, 210*, 115–125. https://doi.org/10.2495/ARC220101

Jiang, B., Song, Y., Li, H. X., Lau, S. S. Y., & Lei, Q. (2020). Incorporating biophilic criteria into green building rating tools: Case study of Green Mark and LEED. *Environmental Impact Assessment Review, 82*(September 2019). https://doi.org/10.1016/j.eiar.2020.106380

Kellert, S., & Calabrese, E. (2015). The practice of biophilic design. *London: Terrapin Bright LLC, 3*(21).

Martin, A. J., Mansour, M., & Malmberg, L. E. (2020). What factors influence students' real-time motivation and engagement? An experience sampling study of high school students using mobile technology. *Educational Psychology, 40*(9), 1113–1135.

Moslehian, A. S., Roös, P. B., Gaekwad, J. S., & Van Galen, L. (2023). Potential risks and beneficial impacts of using indoor plants in the biophilic design of healthcare facilities: A scoping review. *Building and Environment, 233*, 110057.

Nitu, M. A., Gocer, O., Wijesooriya, N., Vijapur, D., & Candido, C. (2022). A biophilic design approach for improved energy performance in retrofitting residential projects. *Sustainability (Switzerland), 14*(7). https://doi.org/10.3390/su14073776

Peters, T., & D'Penna, K. (2020). Biophilic design for restorative university learning environments: A critical review of literature and design recommendations. *Sustainability, 12*(17), 7064.

Sadick, A. M., Kamardeen, I., & Vu, X. P. (2023, September). Challenges for implementing biophilic strategies in Australian building design. *Journal of Building Engineering, 74*, 106849. https://doi.org/10.1016/j.jobe.2023.106849

Sevinc Kayihan, K., Ozcelik Guney, S., & Unal, F. C. (2018). Biophilia as the main design question in architectural design studio teaching. *Megaron, 13*(1).

Söderlund, J., & Newman, P. (2015). *Biophilic architecture : A review of the rationale and outcomes*, 2(May), 950–969. https://doi.org/10.3934/environsci.2015.4.950

Urban Redevelopment Authority. (n.d.). Urban redevelopment authority. https://www.ura.gov.sg/Corporate/Guidelines/DevelopmentControl/NonResidential/Commercial/Advisory-Notes

William Browning, D., Ryan, C. O., & Clancy, J. O. (2014). *14 patterns of biophilic design*. Terrapin Bright Green LLC.

Wilson, E. (1992). *Biophilia* (pp. 79). Harvard University Press.

Yusufov, M., Nicoloro-SantaBarbara, J., Grey, N. E., Moyer, A., & Lobel, M. (2019). Meta-analytic evaluation of stress reduction interventions for undergraduate and graduate students. *International Journal of Stress Management, 26*(2), 132.

Zhong, W., Schröder, T., & Bekkering, J. (2022). Biophilic design in architecture and its contributions to health, well-being, and sustainability: A critical review. *Frontiers of Architectural Research, 11*(1), 114–141. https://doi.org/10.1016/j.foar.2021.07.006

Assessing the Performance of Vernacular and Green Buildings in a Warm and Humid Climate

Niharika Barwal and J. Vijayalaxmi

Abstract As urbanization accelerates, particularly in developing countries like India, the challenge of achieving sustainable cities (SDG Goal 11) becomes increasingly urgent. Currently, green building ratings serve as essential tools for assessing sustainability; however, their application remains limited to large-scale commercial projects within India. Meanwhile, the rich traditional knowledge embedded in Indian vernacular architecture, which has promoted sustainability for centuries, is gradually fading. Despite their differing approaches to sustainability, both green building ratings and traditional architectural practices offer valuable insights for developing a foundation for sustainable residential design in India. This paper explores the potential for integrating green building principles with traditional sustainable practices, focusing on residential construction in Chennai. By examining key differences and commonalities between these two approaches, the research highlights comparative design, material selection, energy efficiency, and climate responsiveness. It was found that the vernacular residence exhibited a maximum temperature difference of 10.4 °C from the outdoors, whereas the maximum temperature difference recorded for the Green Residence was 6.2 °C. On the other hand, the green building demonstrated an energy saving of over 50% more than the Vernacular Building, based on utility bills from October to December 2023. Each of these advantages presents an opportunity to bridge the gaps and form an integrated approach to sustainable housing. Furthermore, an online questionnaire survey on user preferences indicates strong public support for combining traditional and green building elements, with 71.4% of respondents favoring this approach. The findings underscore the importance of raising awareness about both green building standards and traditional practices, advocating for future research to create culturally sensitive and environmentally responsible strategies that leverage the strengths of both systems. This approach not only addresses the urgent housing crisis—estimated at a shortfall of 19 million units in India (Ministry of Housing and Urban Affairs in Report of Technical Group on Urban

---

N. Barwal (✉)
Sustainable Design Analyst, Kapadia and Associates, Mumbai, India
e-mail: niharikabarwal24@gmail.com

J. Vijayalaxmi
School of Planning and Architecture, Vijayawada, India

J. Vijayalaxmi (ed.), *Sustainable Building Performance*,
https://doi.org/10.1007/978-981-96-2428-7_8

housing shortage (TG12) 2012–17. Ministry of Housing and Urban poverty Alleviation, Government of India, 2017)—but also contributes to achieving the Sustainable Development Goals (SDGs) within the country.

Keywords Green rating · Housing · Indigenous · SDGs · Vernacular

Abbreviations

AAC	Autoclaved Aerated Concrete
BREEAM	Building Research Establishment Environmental Assessment Method
GBRS	Green Building Rating System
GRIHA	Green Rating for Integrated Habitat Assessment
IGBC	Indian Green Building Council
LEED	Leadership in Energy and Environmental Design
USGBC	United States Green Building Council
SDG	Sustainable Development Goals

1 Introduction

Buildings have a profound environmental impact, consuming resources such as trees, soil, water, and energy. The primary goal of green buildings is resource efficiency, reducing non-renewable demand, and promoting recycling, reuse, and renewable resource utilization (Makandar & Sanadi, 2019). Green ratings assess buildings based on their life cycle performance, emphasizing the need for credible sustainability proof and measurement for effective management (TERI, 2020).

The modern green building industry's general perception is that green designs incur additional costs and complexities. Contrary to this perception, vernacular communities have historically employed indigenous wisdom, local materials, and sustainable techniques economically. However, these practices are fading with evolving construction methods (Naik, 2016).

1.1 Aim

The research aims to assess traditional vernacular architecture and green buildings using green rating and user satisfaction factors, analyzing the potential for an integrated approach for sustainable housing in warm and humid climates.

1.2 Need for the Study

This study is essential to address the limited application of green building principles in small-scale Indian contexts, the diminishing traditional knowledge in architecture, and the urgent need for sustainable residential construction. By integrating green criteria with traditional practices, the study aims to bridge gaps and meet the demand for climate-responsive and culturally sensitive solutions in residential developments to achieve SDGs in India.

1.3 Background

1.3.1 Traditional Building Systems

Traditional architecture has resulted from generations of survival habits under the constraints of available resources. It was developed over generations and responded to climate change, natural disasters, and local culture. The indigenous architecture has a range of sustainable practices in all its climatic contexts (Vijayalaxmi & Arathy, 2022).

The concept of home is a central reflection of various social and cultural values, shaping individual and family identities. Residences carry psychological, social, and cultural significance, becoming repositories where cultural elements assimilate and acquire diverse meanings and memories for occupants. Housing spaces, influenced by cultural components and dimensions, play a crucial role in meeting societal and climatic requirements. Vernacular architecture, responsive to local needs and cultural demands, is complex due to diverse spatial forms and layouts.

Each housing space is associated with habitual, cultural, religious, social, and economic activities, creating episodic memories tied to emotions. A study by Komagal and Subramanian (2022) examines material construction, form variations, and overt behaviors among occupants, including positive thinking, affection, empathy, comfort, composure, and control in selected traditional home settings in Tamil Nadu. House front sit-out, or Thinnai, is highlighted for its immense significance as the first space to interact with visitors and invite them into the home. Some of the physical features with traditional socio-cultural values of Tamil vernacular architecture are:

- Front sit-out (Thinnai)
- Central courtyard (Mutram)
- Space around the courtyard (Sutrukattu)
- Backyard (Kollaipuram).

1.3.2 Green Building Rating Systems (GBRS)

Building sustainability and greenness are assessed and verified using the GBRS, a holistic outline created by construction establishments, private consultancy firms, or international organizations (Hwang & Ming, 2018). It consists of a set of clear performance requirements that a building should achieve to be certified, along with specific recommendations that could assist the project crew in meeting or exceeding their performance requirements (Mattoni et al., 2018). According to Kubba (2010), and Pearce and Ahn (2012), GBRS describes how environmentally friendly and ecologically sound a structure is, shows the amount to which green features have been implemented, and indicates the sustainable concepts and practices that have been used. Some of the Indian Green Building Rating Systems are introduced below:

1. Indian Green Building Council: IGBC is initiated by the Confederation of Indian Industry (CII), responsible for promoting green practices in India. It offers several green building rating manuals, including IGBC Green Townships, IGBC Green Homes, IGBC Green Factories, etc.
2. Green Rating for Integrated Habitat Assessment: GRIHA, an indigenous GBRS, was established by TERI (The Energy and Resources Institute), India. GRIHA ratings range from one star (minimum compliance) to five stars (highest sustainability performance).
3. Leadership in Energy and Environmental Design: The USGBC introduced LEED, a globally recognized green building certification system. Buildings may obtain Certified, Silver, Gold, or Platinum certification levels.

1.3.3 Literature Review

The global building industry greatly impacts global environmental and energy consumption issues. There are both direct and indirect requirements for energy in buildings due to the consumption throughout their lifespan (Alrashed et al., 2017). As a result of this issue, efficiency, and thermal comfort have emerged as critical concerns in the decision-making and design phases of green and sustainable buildings (Agyekum et al., 2020). Several building rating systems are created to be used throughout the world, including BREEAM (UK/Europe), Leadership in Energy and Environmental Design (LEED; the USA, affiliated worldwide), Green Star (Australia), etc. GRIHA (Green Rating for Integrated Habitat Assessment) and IGBC (Indian Green Building Council) are two programs that have been designed for sustainable development throughout India. They aim to develop more efficient urban forms by utilizing superior engineering, design, and planning (Smith, 2015).

However, the rating systems have numerous limitations. According to Makandar & Sanadi (2019), LEED and GRIHA rating systems are only appropriate for high-cost construction, and large projects. These rating systems are unlikely to yield efficient results because their criteria consider metro cities and adjacent areas while ignoring tier-II towns and cities.

The incorporation of local crafts into the green building process has the potential to be a critical solution to addressing the issues of sustainable design in a more economically equitable and socially inclusive manner than what prevalent green building ideologies—often adopted from the West and tweaked for the Indian context—illustrate, implement, and endorse (Naik, 2016). Nagan et al. (2011) studied the traditional architecture of the Chettinadu house for its passive cooling and ventilation techniques and concluded that incorporating such techniques can enhance sustainability by improving energy efficiency and providing thermal comfort.

Even while vernacular resources, technologies, or systems are often sustainable, practical constraints from changing social, cultural, and technical settings could be present. To determine how the knowledge, and skill of global vernacular architecture can be effectively utilized in a modern environment is a challenge, as Asquith and Vellinga (2006) has stated (Wahid, 2012). The study is part of a broader effort to comprehend and analyze the traditional building knowledge from ancient literature and treatises on achieving sustainability. More interventions in architecture are required to explore alternate paths and approaches to transferring traditional knowledge to a future sustainable built environment.

The modern green building industry's general perception is that green designs incur additional costs and complexities; however, vernacular crafts communities have cultivated strategies for building with indigenous wisdom, local materials, and sustainable techniques most economically and equitably for centuries (Naik, 2016). On the contrary, indigenous practices are becoming obsolete with rapidly changing construction techniques. These can be studied with the possibilities of integration for spreading awareness regarding sustainability and providing an Indian context for green building developments.

1.4 Scope and Limitations

The research encompasses a comprehensive examination of both green building rating systems (GBRS) and the traditional design practices in residential construction specific to Chennai, and Tamil Nadu in the Indian context. It will involve case studies of a vernacular structure, and a building aiming for a green rating. A comparative analysis of case studies in the warm and humid climate, considering user perception and green rating criteria (IGBC) will be studied.

The study is limited to residential construction in the warm and humid climate of Chennai, considering the IGBC Residential green rating case study.

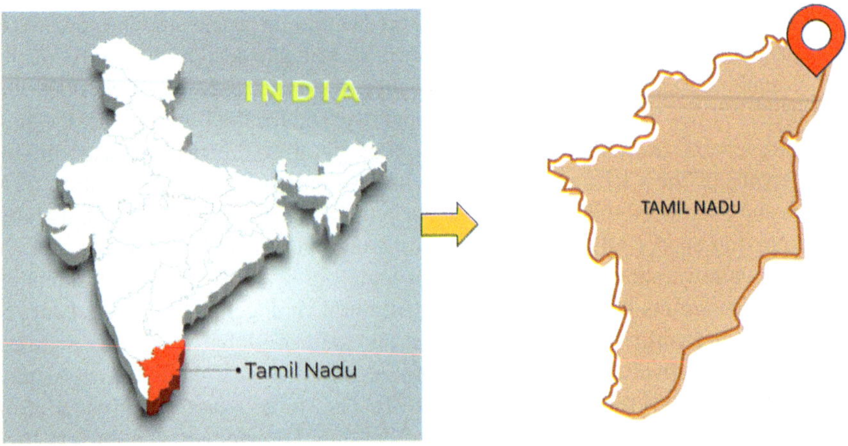

Fig. 1 Location of study area—Chennai. *Source* shutterstock.com/

2 Materials and Methods

2.1 Study Area

Chennai (13.08° N; 80.27° E) is the capital city of Tamil Nadu and the fourth largest city in India situated in the southern part of the country (Fig. 1). It has a warm and humid climate (NBC, 2016). In 2024, the present approximate population of Chennai city is 6,599,000, with the estimated metropolitan population reaching 12,290,000. The current city population estimates rely on historical growth rates. The cultural identity of Chennai, often referred to as the "Gateway to South India," stands out from that of any other Indian city. As a cosmopolitan center, the city's culture mirrors its diverse population, giving rise to a distinctive fusion. Despite its status as a modern metropolis, Chennai maintains traditional elements. The city is known for its widespread appreciation of traditional Tamil Nadu art forms, including music and dance. Chennai boasts a unique cultural amalgamation, featuring traditional and fast foods, and ancient temple architecture juxtaposed with modern high-rises.

2.2 Tools and Techniques

- Document Analysis: Analyzing digital documents, research papers, and architectural plans available online for case studies and background information.
- Literature Review: Accessing digital libraries and academic databases for a comprehensive remote literature review.
- Statistical Comparison: Perform statistical analysis of live case studies.

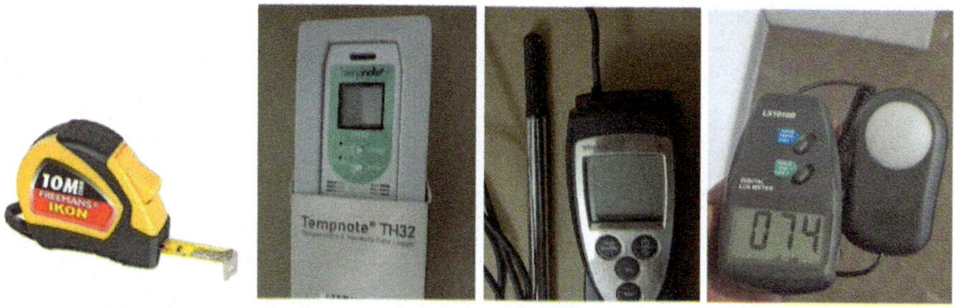

Fig. 2 Tools used. *Source* Author

- Tempnote TH32 was used for the measurements of temperature and humidity indoors and outdoors.
- Site survey and measurements were done using measuring tapes.
- A questionnaire survey was taken from the occupants.
- TESTO 425 (Thermal Anemometer) was used to determine wind speeds indoors.
- Digital lux meter LX1010B used for daylight measurements in the houses.

- Perception Questionnaire: An online survey questionnaire to understand people's perception of using traditional elements and green-rated appliances or criteria in their residences (Fig. 2).

3 Analysis

3.1 Green Residence, Chennai

See Fig. 3.

Location: Madipakkam, Chennai, Tamil Nadu
Built-up area: 388.5 m^2
Climate: Warm and Humid
Typology: Residential
Occupants: 6

Sustainable Design Features:

- Designed in the east–west axis (Fig. 4)
- Passive features include Orientation, Cross ventilation, proper daylighting, and provision of skylight.
- Use of Johnson Cool Roof Tiles (SRI-90).
- Provision of ramp and lift for the elderly.
- Ample parking facility on site.
- Basic household amenities are within walking range of the site.

Fig. 3 Green residence in Chennai

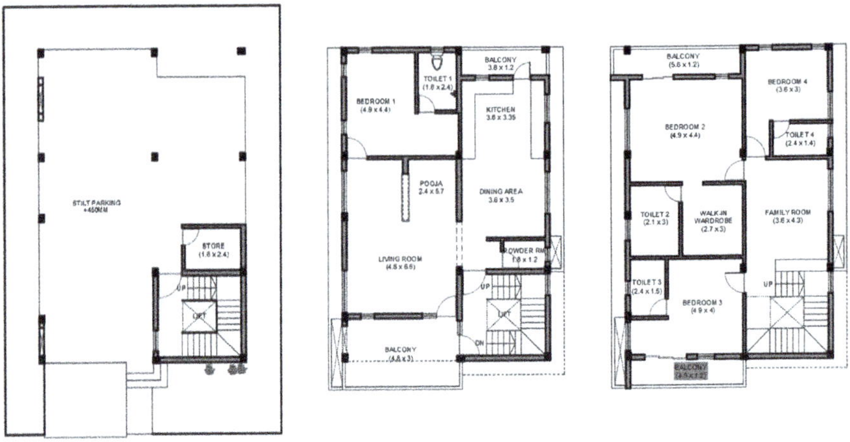

Fig. 4 Floor plans—**a** stilt; **b** first floor; **c** second floor. *Source* Author

Energy Efficiency:

- Recent electricity bills comparison for 01/10/2023–31/12/2023: 405 kWh (1.04 kWh/m^2)
- The temperature and humidity differences from outdoor to indoor as recorded on 12/02/2024–14/02/2024 are given as (Figs. 5 and 6; Table 1):

Materials and Resources:

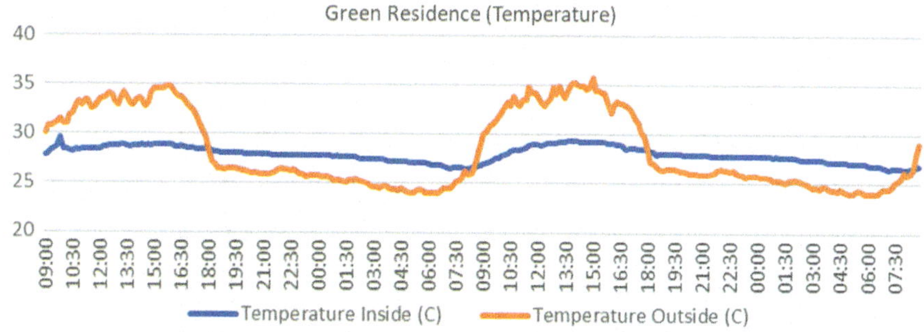

Fig. 5 Measured temperature difference: green building. *Source* Author

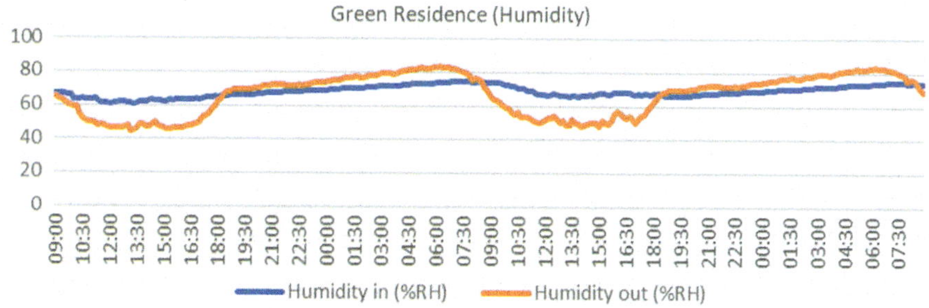

Fig. 6 Measured humidity difference: green building. *Source* Author

Table 1 Measured TH averages—Green Residence

Green	Temp Inside (C)	RH (%)	Temp outside (C)	RH (%)
Max	29.6	74.7	35.8	83.4
Avg	**27.98**	**68.8**	**28.39**	**66.98**
Min	26.6	60.9	24	44.4

Source Author

- Segregation of waste on site.
- Use of recycled content like flyash.
- Use of salvaged doors.
- Use of AAC blocks, and porotherm blocks.
- 200 mm AAC Wall U-value is 0.44 W/m^2 K and time lag is 8.95 h (Figs. 7 and 12).

Resident Health and Well-Being:

The daylight was measured on 12th February 2024 at 9:00 am and 12:00 pm on a clear sunny day using a digital lux meter LX1010B. Recorded daylighting was found to be > 110 lx for the habitable rooms (Table 2).

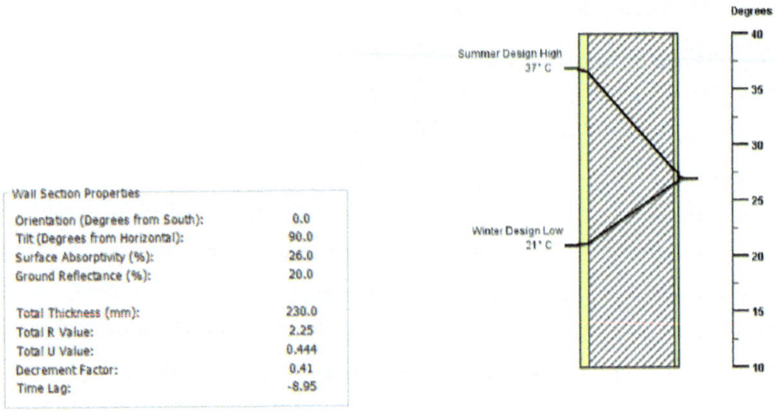

Fig. 7 Wall section—Green Building

Table 2 Illuminance levels—Green Residence

	9 AM	12 PM
Staircase lobby (GF)	240–250 lx	310–320
Drawing room	120–130	140–150
Bedroom 1	160–170	190–200
Bedroom 2	220–230	290–300
Kitchen/Dining	160–170	180–190
Bedroom 3	200–210	230–240
Family room	770–780	840–850
Bedroom 4	150–160	160–150

Source Author

- All the rooms are naturally ventilated, and ACs are used during summer.
- Cross ventilated areas.
- The overall thermal comfort of the users in the house had a mean of 6.18 (Very satisfied).
- The users said that there is a need for air conditioners and coolers for at least 3 h per day during the summer.

3.2 *Vernacular Building at Kanchipuram*

See Fig. 8.

Location: Kanchipuram, Tamil Nadu
Built-up area: 135.5 m^2

Fig. 8 Vernacular residence at Kanchipuram. *Source* Author

Climate: Hot and Humid
Typology: Residential
Occupants: 6

Sustainable Design Features:

- Strictly follows the east–west axis (Fig. 14).
- Use of sloping roof with Mangalore tiles sloping roof (SRI-36).
- Passive features include orientation, use of thermal mass, courtyard integration, and small skylights on the roof.
- No universal design provision. Hindering levels.
- No parking facility on site.
- It is on the outskirts, so amenities are not easily accessible (Fig. 9).

Energy Efficiency:

- Recent electricity bills comparison for 01/10/2024–31/12/2024: 305 kWh (2.25 kWh/m^2)
- The temperature and humidity differences from outdoor to indoor as recorded on 13/02/2024–15/02/2024 are given as (Figs. 10 and 11; Table 3):

Materials and Resources:

- Waste is segregated and kitchen peels are fed to the cattle.
- Use of natural materials.
- Later repair work was done with concrete flooring and brick columns in the courtyard.
- 400 mm thick mud brick wall with U-value of 1.9 W/m^2 K and time lag of 11.49 h (Fig. 12).

Fig. 9 Plan-vernacular residence n.t.s. *Source* Author

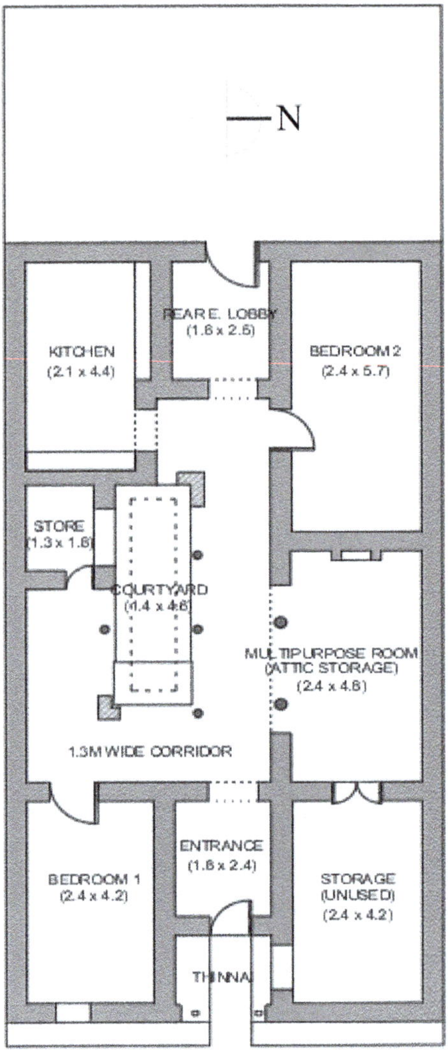

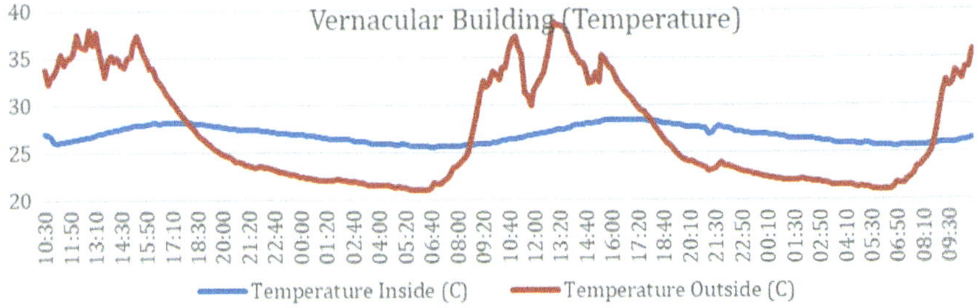

Fig. 10 Measured temperature difference—vernacular building. *Source* Author

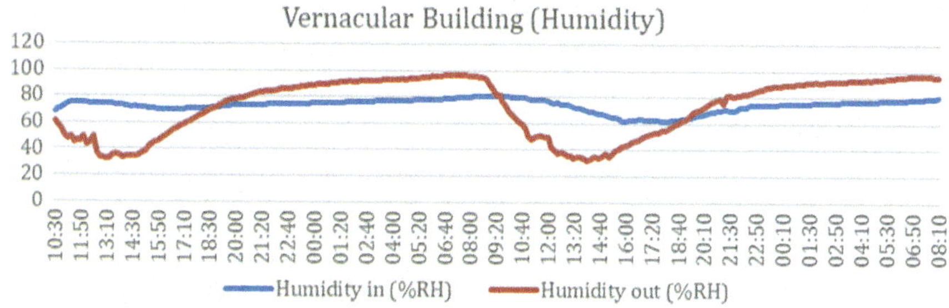

Fig. 11 Measured humidity difference—vernacular building. *Source* Author

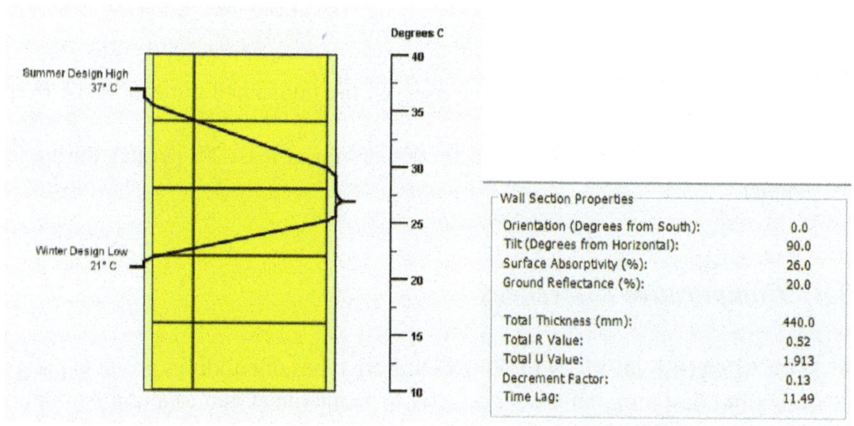

Fig. 12 Wall section—vernacular building

Table 3 Measured TH averages—vernacular residence

Vernacular	Temp Inside (C)	RH in (%)	Temp Outside (C)	RH out (%)
Max	28.4	80.7	38.8	96.8
Avg	**26.89**	**74.18**	**27.11**	**73.09**
Min	25.6	62.3	21	32.3

Source Author

Resident Health and Well-Being:

The daylight was measured on 13th February 2024 at 9:00 am and 12:00 pm on a clear sunny day using a digital lux meter. It was found to be less in the rooms as the courtyard and tile slit skylight was the only source of light in the building (Table 4).

- All the rooms are naturally ventilated opening toward the courtyard.
- No cross ventilation.

Table 4 Illuminance levels—vernacular residence

	9 AM	12 PM
Thinnai	360–370	320–330
Entrance	60–70	70–80
Bedroom 1	50–60	40–50
Bedroom 2	100–110	130–140
Corridor	940–950	1180–1190
Kitchen	40–50	70–80
Courtyard	39,090–39,100	47,600–50,000
Multipurpose room	130–140	180–190

Source Author

- The overall thermal comfort of the users in the house had a mean of 6.4 (Very satisfied).
- The users said that there is a need for coolers for at least 2 h per day during the summer.

3.3 Comparative Inferences

- SUSTAINABLE DESIGN FEATURES: Both types of buildings promote passive designing, designing an east–west axis to reduce heat gain, natural ventilation strategies, etc. While the universal design and modern amenities are required but lacking in the vernacular house.
- ENERGY EFFICIENCY: The kWh/m^2 electricity consumption was higher in the vernacular house due to less efficient appliances. Upon comparison of outdoor to indoor temperatures, both the houses performed equally well, considering the indoor temperatures lie within the comfort range. At the same time, the scope of the study was limited to better comfort season during the February field study, field measurements of extreme summer months would give better clarity on the envelope performance.
- MATERIALS and RESOURCES: The u-value of the AAC blocks was found to be 0.44 W/m^2 K offering better thermal insulation as compared to 1.9 W/m^2 K of a mud wall. However, mud wall has a higher thickness, and thermal lag, and provide good natural cooling through thermal mass, absorbing heat during the day and releasing it at night.

Also, the cool roof performs better however the cost and relative performance of sloping roofs of a vernacular house are better. The use of natural materials can help reduce the cost of the green building, using low maintenance natural materials.

- RESIDENT HEALTH and WELL-BEING: While the thermal comfort of the occupants was satisfactory, the daylight of the vernacular house was significantly lower. This reduction in openings resulted in better thermal comfort for the occupants but with the changing lifestyles, significant daylighting is required indoors.
- USER SATISFACTION: A survey of the six occupants was taken in both residences on a seven-point scale (1-Very Dissatisfied; 2-Dissatisfied; 3-Slightly Dissatisfied; 4-Neutral; 5-Slightly Satisfied; 6-Satisfied; 7-Very Satisfied). The results of the user satisfaction survey are given below (Table 5):

- The users of the Green Residence were neutral about their satisfaction levels with the outdoor noise levels, which is possible due to closely spaced development and lack of nature integration as noise barriers.
- The vernacular house, on the other hand, is less functional according to the users as it cannot be used multi-purposely with the changing lifestyles of the users and the integration of modern amenities.
- The suitability of the vernacular house during COVID-19 living conditions was found slightly dissatisfactory for the users.
- The vernacular house users marked 7 for the amount of parking area, although there is none present on site, however, there are ample unauthorized spaces to park in the village.
- USER SATISFACTION-PSYCHOLOGICAL: The green building users responded that their house suits their way of life (7) and they are satisfied with how the long hours spent indoors affect their mood positively (6.5). They responded to the impact of technology (TVs/Phones) in their daily life and interactions sometimes (3.5). On the other hand, the vernacular house occupants were neutral (4.4) about their house suiting their way of life, they mentioned more insect infestation due to the presence of a courtyard. The users responded to the impact of technology (TVs/Phones) in their daily lives and interactions as never (1).
- SOCIO-CULTURAL FACTORS (Table 6):

Table 5 User satisfaction based on functionality

Questions on satisfaction levels based on function	**Green**	**Vernacular**
The overall comfort of living at your house?	5.966	5.292
The appearance of your house?	6.382	5.916
The privacy of the rooms in your house?	5.674	6.481
The sizes and sufficiency of rooms in your house?	6.069	5.000
The number of parking areas in your house?	7.000	7.000
The suitability for your housing needs during the COVID-19 lockdown?	7.000	3.000
The outdoor noise levels entering your house?	4.939	6.481
The accessibility of your house to public and social amenities? (School, shopping, transport, etc.)	6.581	4.899

Source Author

Table 6 User satisfaction based on socio-cultural aspects

Questions	Green	Vernacular
How well does the design support various family structures and dynamics, considering different generations or extended family living arrangements?	6.58	3.464
How often are you involved in the community activities around your house?	1.38	4.472
What kind of social activities do you participate in?	Rarely	Festival/Comm. Gath
Do you hold any social activities/gatherings at your residence during festivals/other occasions?	6.153	4.000
How often do your neighbors visit/talk?	3.201	3.873
How satisfied are you with the spaces for social interactions around your house?	6.382	6.000
Are there spaces that encourage conversations in the house?	Y	Y
Does the use of local materials or any features contribute to a sense of connection to the broader community?	Y	Y
How often do you visit your village/native place? (if any)	2.048	3.464

- However, green building users rarely participate in the community activities around the house, they mentioned various preferences like cleanliness drives, stray animal welfare, etc.
- Only the elderly in the house often have time to converse with the neighbors.
- The vernacular house users often participate in the community activities around the house mostly during festivals and community gatherings.
- The interactions with the neighbors have seemed to be decreasing over the years for the vernacular house users.
- A study by Komagal and Subramanian (2022) examines material construction, form variations, and overt behaviors among occupants, including positive thinking, affection, empathy, comforting, composure, and control in selected vernacular home settings. House front sit-out, or *Thinnai*, is highlighted for its immense significance as the first space to interact with visitors and invite them into the home. Some of the physical features with traditional socio-cultural values of Tamil vernacular architecture are the front sit-out (*Thinnai*); Central courtyard (*Mutram*); Space around the courtyard (*Sutrukattu*); and Backyard (*Kollaipuram*).
- The integration of socio-cultural elements in the house is important as can be seen in the vernacular house with the increased interactions with the community, however, the willingness to contribute to a larger goal in a community by Green Residence users can also have a positive impact. These are various factors that shouldn't be neglected in the integration of green and traditional buildings to create a uniquely bound sustainable system (efficient, social, and environmentally sound) in an Indian context.

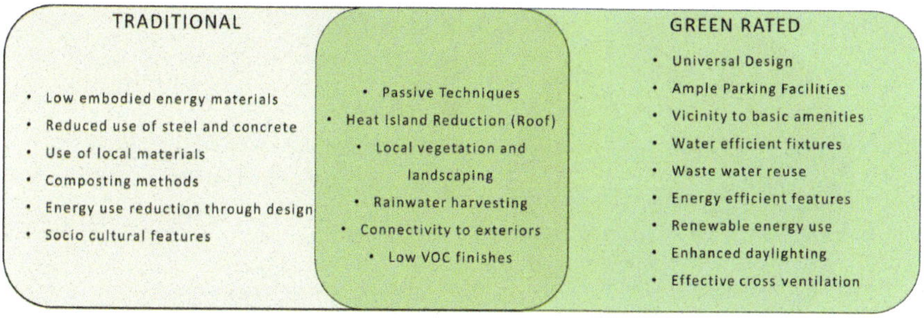

Fig. 13 Common and distinct parameters

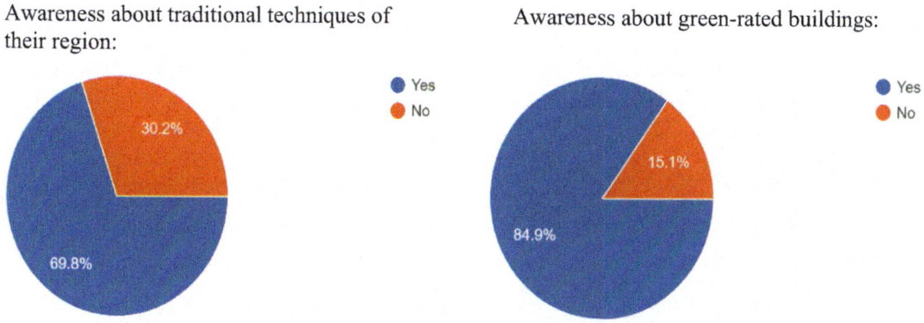

Fig. 14 People awareness: traditional architectural techniques and green-rated buildings

From the studied aspects of both approaches, the following were found to be major common and distinct sustainable features of traditional and green-rated methods (Fig. 13).

3.4 User Preference Questionnaire

To understand the awareness and preferences of Indian users regarding the vernacular and green building elements, an online questionnaire survey was conducted. The survey has 106 respondents, some relevant responses are provided as follows (Figs. 14, 15, 16 and 17):

4 Results

This comparative study of the Green Residence in Chennai and the Vernacular Building in Kanchipuram reveals insights into the interplay between modern sustainable practices and traditional architectural knowledge in the context of warm and

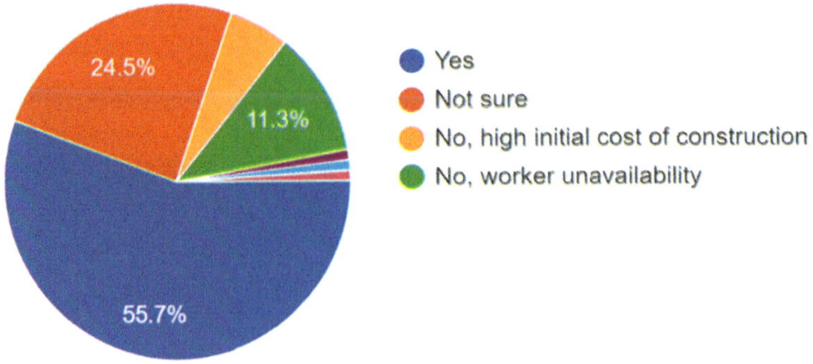

Fig. 15 Preference for living in a traditional/vernacular building

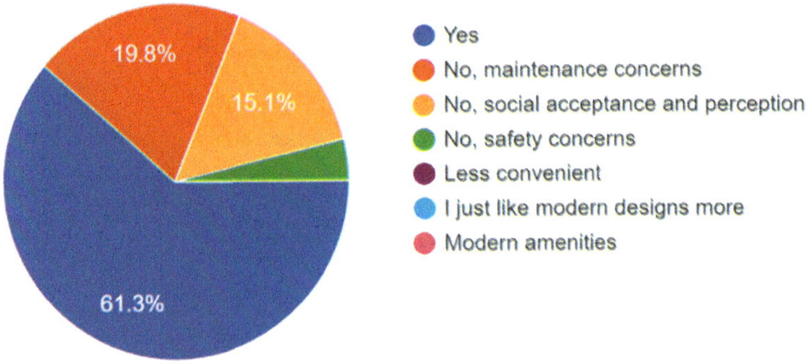

Fig. 16 Preference for living in a green-rated building

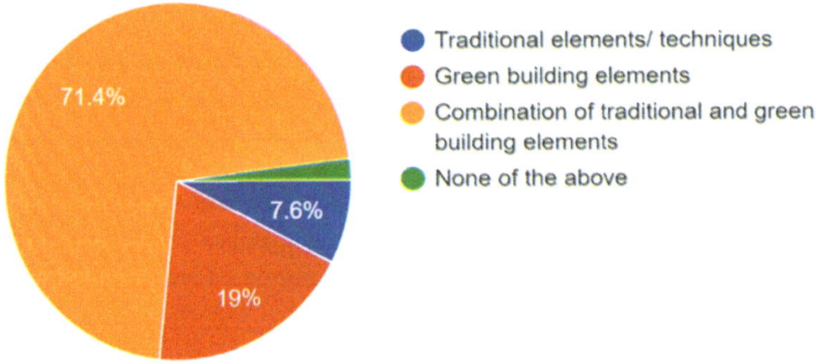

Fig. 17 Preference of adopting elements at their own residence

humid climates. Both building typologies successfully employ passive design strategies, such as alignment along the east–west axis and natural ventilation, to enhance thermal comfort and energy efficiency. However, notable differences emerge in their functionality and user satisfaction. The Green Residence demonstrates superior energy efficiency with over 50% lower kWh/m^2 consumption. On the other hand, the Vernacular Building, with its higher thermal mass and traditional materials, experiences a higher temperature difference of 10.4 °C than 6.2 °C in the green building. Daylighting levels in the Vernacular Building are notably lower, which may compromise the residents' well-being, highlighting the need for enhanced lighting solutions in Vernacular Buildings to adapt to contemporary living standards. An integrated checklist of features is shown in the comparison of both approaches (Fig. 13) to be applied for an integrated sustainable approach to housing design.

5 Discussion

User satisfaction surveys indicate that the Green Residence meets the demands of modern lifestyles more effectively, with higher ratings for functionality and comfort. Residents appreciate the integration of contemporary features that support social interaction and accessibility. Conversely, while the Vernacular Building fosters community engagement and cultural continuity, it falls short in accommodating the evolving needs of its occupants.

The questionnaire from over 100 people indicates a strong inclination of the respondents (71.4%) toward a fusion of traditional and green building elements. As the study was only limited to a few techniques particular to Chennai, proper documentation of diminishing traditional practices has to be done for effectively implementing these techniques for green buildings in different contexts in India. A breakdown of various techniques and their benefits for building performance in various climates is important for its implementation amalgamating these sustainable approaches to a common goal for sustainable development. The feasibility of these techniques on a larger scale and high-rise residences also needs to be looked upon for an integrated design.

6 Conclusion

The study emphasizes how traditional and green building approaches have a positive impact on both of their performance. It underscores the potential for a synergistic approach that combines the strengths of both green building principles and traditional design elements. By integrating efficient technologies with the rich cultural context of vernacular architecture, future residential projects can create environments that are not only energy-efficient but also culturally resonant and socially supportive. This integration is essential for developing sustainable urban settlements in India, addressing both the immediate challenges of urbanization and the long-term goals of SDG 11.

References

Agyekum, K., Danku, J. C., & Kissi, E. (2020). Professionals' views of vernacular building materials and techniques for green building delivery in Ghana. *Scientific African, 8*(2020). https://doi.org/10.1016/j.sciaf.2020.e00424

Alrashed, F., Asif, M., & Burek, S. (2017). The role of vernacular construction techniques and materials for developing zero-energy homes in various desert climates. *Buildings, 7*(2017), 1–19. https://doi.org/10.3390/buildings7010017

Asquith, L., & Vellinga, M. (2006). *Vernacular architecture in the 21st century* (pp. 111–119). Taylor & Francis Group. https://doi.org/10.4324/9780203003862

https://www.grihaindia.org/

https://www.teriin.org/

https://igbc.in/

Hwang, B.-G., & Ming, S. (2018). Green building rating systems: Global reviews of practices and research efforts. *Sustainable Cities and Society*.https://doi.org/10.1016/j.scs.2018.02.034

Komagal, A., & Subramanian, C. V. (2022). Spatial analysis on psychodynamic influences on dwelling spaces of vernacular architecture—Tamil Nadu. *Journal of Algebraic Statistics, 13*(3), 2783–2797. https://publishoa.com ISSN: 1309-34522783

Kubba, S. (2010). *Green construction project management and cost oversight*. Elsevier.

Makandar, S. S., & Sanadi, N. A. (2019). Limitations of green building rating systems—A case of LEED and GRIHA. *International Research Journal of Engineering and Technology, 6*(5). Retrieved from: https://www.irjet.net/

Mattoni, B., Guattari, C., Evangelisti, L., Bisegna, F., Gori, P., & Asdrubali, F. (2018). Critical review and methodological approach to evaluate the differences among international green building rating tools. *Renewable and Sustainable Energy Reviews, 82*, 950–960. https://doi.org/10.1016/j.rser.2017.09.105

MoHUA. (2017). Report of Technical Group on Urban housing shortage (TG12) 2012–17, Ministry of Housing and Urban poverty Alleviation, Government of India. Accessed as https://www.mohua.gov.in/upload/uploadfiles/files/1Ministry_HUPA_Annual_Report_2016_17_English_small$2017Mar17170100.pdf

Nagan, S., Radhakrishnan, S., Shanthi Priya, R., & Sundarraja M. C. (2011). Climate responsive traditional architecture of Chettinadu Housing in Tamilnadu, India—A qualitative and quantitative analysis during summer. *International Journal of Ventilation, 10*(1). 89–97.https://doi.org/10.1080/14733315.2011.11683937

Naik, A. (2016). *Advancing equity and inclusion in Green building design in India*. Insight, September 2013. Retrieved from: https://www.arthastudio.com/

NBC. (2016). SP 7 : 2016, National Building Code of India 2016. Accesses on 24th March 2025 at https://www.bis.gov.in/standards/technicaldepartment/national-building-code/

Pearce, A., & Ahn, Y. H. (2012). *Hanmi global sustainable buildings and infrastructure*. New York: Routledge.

Smith, R. M. (2015). "Green" building in India: A comparative and spatial analysis of the LEED-India and GRIHA rating systems. *Asian Geographer*. https://doi.org/10.1080/10225706.2015.1020065

TERI. (2020). Why green ratings for buildings matter?. 18 Feb 2020 Mr Santhosh Ramkumar. Accessed on 24th March 2025 on https://www.teriin.org/article/why-green-ratings-buildings-matter

Vijayalaxmi, J., & Arathy, K. C. K. (2022). Critical evaluation of socio-cultural and climatic aspects in a traditional community: A case study of Pillayarpalayam weavers' cluster, Kanchipuram. *Built Heritage, 6*, 3. https://doi.org/10.1186/s43238-022-00050-4

Wahid, A. (2012). Adaptive vernacular options for sustainable architecture. *Journal of the International Society for the Study of Vernacular Settlements, 2*(2), 74–87. Retrieved from: https://isvshome.com/

A Study on Optimisation of Wind-Flow in Residential Complexes for Chennai

M. Neha Shremitha and J. Vijayalaxmi

Abstract Sustainable Architecture has become increasingly relevant in the context of global warming. The rising population contributes to higher energy consumption and demand, which in turn leads to increased greenhouse gas emissions and climate change. A significant method for reducing energy use in buildings is through creating thermal comfort passively via Natural Ventilation; particularly in Hot-Humid regions. This paper seeks to identify an effective building block orientation and arrangement within a residential complex in Chennai, utilising three key site parameters: dimensions, orientation, and spacing between blocks; employing Computational Fluid Dynamics (CFD) simulations to evaluate how these factors influence indoor wind comfort indices. Final findings indicate that Staggered block arrangement with an orientation angled between + 30° and + 45° towards the South-East is optimal for capturing the annual winds (South-West), achieving ideal wind velocities. Furthermore, the linear interior layout performs superior compared to other proposed designs. Also, the study emphasises the analysis of interior layouts in relation to the identified optimal block configuration, serving as a foundational guide for designing building block arrangements that promote sustainable housing through wind-driven Natural Ventilation.

Keywords CFD · Housing · Natural ventilation · Passive cooling

M. Neha Shremitha (✉)
Sustainable Architect, ENGRIN, Hyderabad, India
e-mail: ar.nehashremitha.maheswaran@gmail.com

J. Vijayalaxmi
School of Planning and Architecture, Vijayawada, India

J. Vijayalaxmi (ed.), *Sustainable Building Performance*,
https://doi.org/10.1007/978-981-96-2428-7_9

1 Introduction

Sustainable Housing has become important in today's global scenario to prevent climate change brought on by global warming. Buildings consume over 40% of global energy and emit greenhouse gases. In India, buildings alone consume one-third of India's total energy. Sustainable housing ensures people's well-being and also has a lower environmental impact (BEE, 2020), hence utilising renewable and clean energy sources, such as wind, is a crucial passive strategy for combating global warming. (Abbasi, 2021; Zhou et al., 2014). Most people spend a large amount of time at home, so it is important to design buildings that are both functional and thermally comfortable achieved by using renewable energy sources (Daemei et al., 2016).

It is established that comfort ventilation is the most effective passive strategy to enhance indoor comfort in hot-humid climates (Vijayalaxmi, 2013). Also it helps in reducing the physiological effect of high humidity in hot-humid climate with less diurnal temperature ranges (Vijayalaxmi, 2009). Natural ventilation has been studied extensively using CFD simulation methods to examine the effects of natural ventilation in both outdoor and interior thermal conditions by previous researchers. This approach has proven effective in the early stages of concept design (Marija, 2007) with discrepancy ranging from 0.9 to 7.8%, when compared with the results of mathematical Network model (Asfour, 2010). Later, the study expanded to look onto natural ventilation for indoor and outdoor buildings on a larger scale, with practical wind tunnel methods validating their numerical simulation approaches (Yang et al., 2019).

Studies on optimising natural ventilation using building parameters like: form, orientation, and building intervals at site level are rarely studied for housing complexes, especially in Hot-Humid climate of Chennai, India. Consequently, this study will benefit in enhancing indoor thermal performance by passive comfort ventilation through optimal building form configuration at initial design level.

2 Methodology

The flow of this research method is shown in Fig. 1. The initial climatic and wind data like wind velocity and direction for analysis is collected from Climate Consultant software tool version 6. The predominant annual wind direction for Chennai is South-West, as studied from the Climate Consultants annual wind rose charts. Many factors like site location, climate, building block dimensions, shape and orientation influence the building's natural ventilation. However, in this study contextual elements like buildings and trees were not included in the simulation done by CFD tool in Design Builder (version 7.0.1.006).

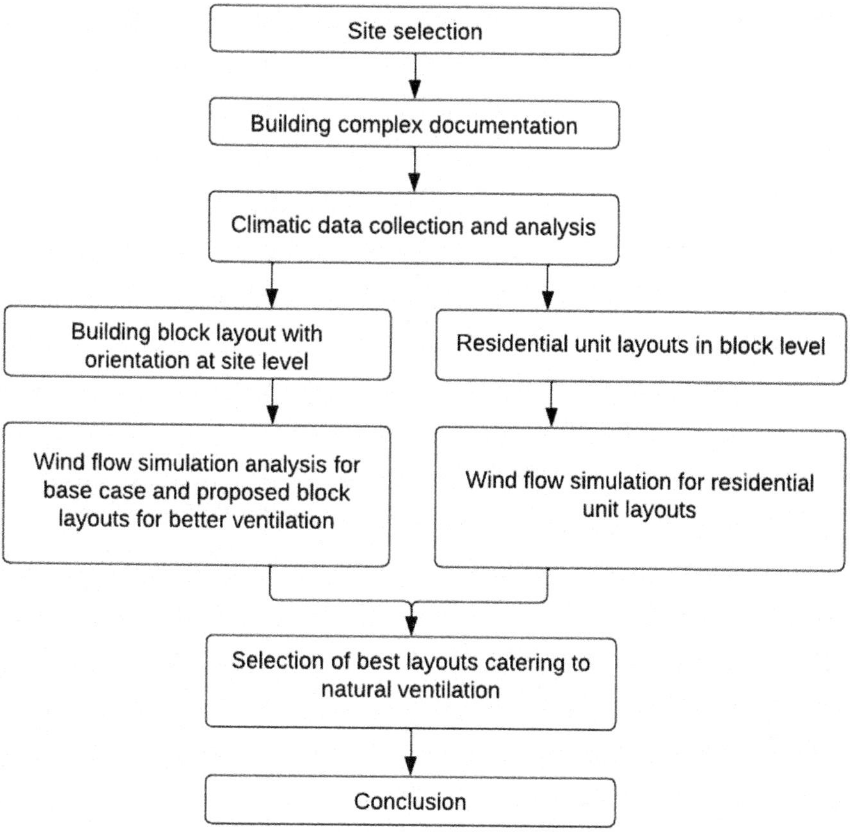

Fig. 1 Research method process

2.1 *The Site*

The considered residential complex for the study is located in Sholinganallur, Chennai as depicted in Fig. 2, developed by Tamil Nadu Housing Board (TNHB) comprising 36 blocks categorised by economic income groups: Low Income (LIG), Middle Income (MIG), and High Income (HIG). The complex features 160 residential units: 40 HIG, 60 MIG, and 160 LIG units. The existing conditions, outlined in Table 1, are simulated to evaluate natural ventilation effectiveness and compared with alternative block orientations of the same density. The proposed blocks are simulated for 5 different orientations in the site as shown in Table 2.

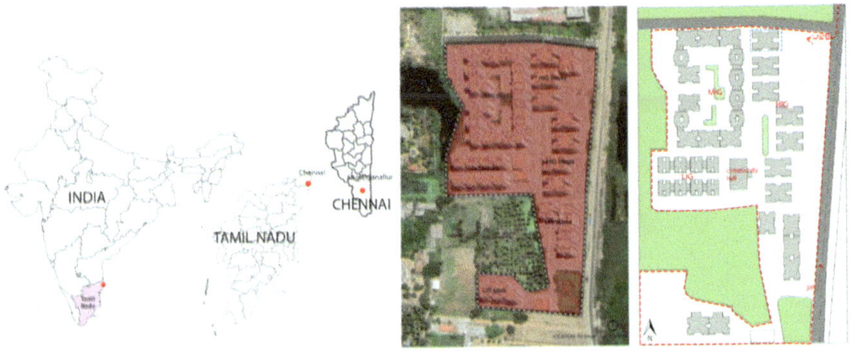

Fig. 2 Selected site location map in Chennai

Table 1 Existing characteristics of the TNHB complex

Name of the block	Block dimensions (length × breadth) in metres	Number of floors	Number of blocks	Site orientation
Existing				– 5°SW
HIG	31 × 21	10	8	
MIG	25 × 19	10	14	
LIG	14 × 18	10	14	
Community centre	21 × 35	2	1	

2.2 *CFD Simulation Inputs and Method*

The input data for Design Builder CFD is given in Table 3. There are different conditions used by different countries to categorise the comfort conditions. With regards to the wind, velocity is one of the criteria to establish comfortable indoor condition. The comfort index for indoor wind conditions is shown in Table 5, whereas Table 4 shows human perception with respect to different wind velocities (Egan, 1975; Esfahankalateh et al., 2021; Hong & Lin, 2015).

The simulations were conducted as per the boundary parameters displayed in Table 3, which was validated by Asfour (2010), who analysed the behaviour of wind movement with different residential block configurations initially in 2010. The existing residential blocks are rearranged with similar size blocks as depicted in Fig. 3, for the sake of analysing the influence of orientation of building forms with wind direction.

The existing residential blocks, along with proposed block layouts, are simulated to identify the optimal block orientation. Table 2. Case B and Case C feature distinct patterns and orientations, with block spacing ranging from 20 to 38 m in accordance with Tamil Nadu building byelaws. Although all blocks have the same size, Cases B and C utilise a detached configuration. Using the determined optimal orientation,

Table 2 Different layout alternatives of block orientation with respect to the site

BUILDING BLOCKS		**ORIENTATION**				
Criteria	**Alternatives**	**-20°**	**-10°**	**0°**	**30°**	**45°**
Linear block arrangement with varying length	Case A			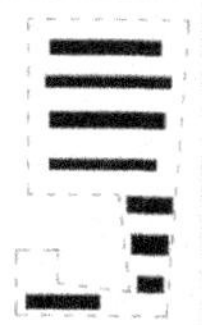		
Scattered pattern block arrange-ment	Case B	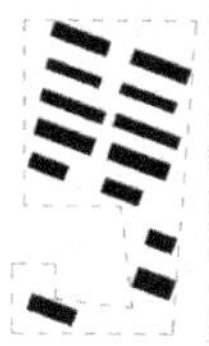	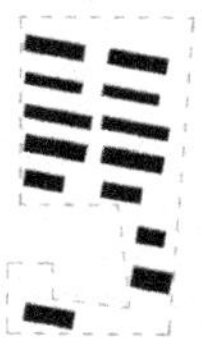	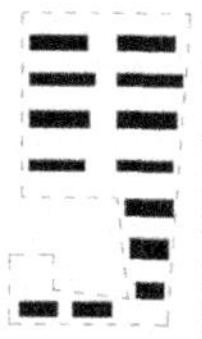		
Stagger-ed pattern block arrange-ment	Case C			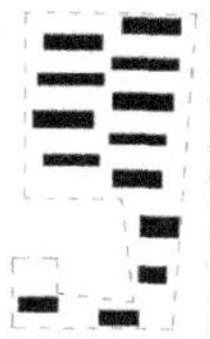	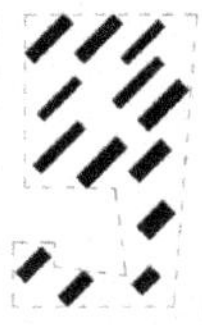	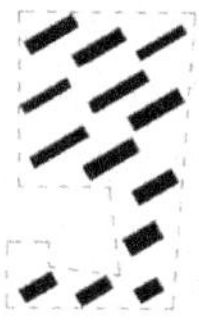

three alternative interior layouts for HIG units are subjected to internal CFD simulations to assess and improve indoor thermal comfort. This approach can also be applied to MIG and LIG units using similar CFD analysis methods (Fig. 4).

3 Results

External CFD simulations were conducted for multiple block configurations, as shown in Table 6, using a CFD plane at a height of 10 m to determine the optimal block orientation. Simulations ensured that all configurations achieved outdoor wind velocities between 2.5 and 4 m/s, a range suitable for outdoor comfort, with block spacing ranging from 20 to 32 m.

Case A, Case B, and Case C in Table 6 explored different block patterns. Case C, with a staggered arrangement, demonstrated superior performance, producing wind velocities between 4.6 and 6.5 m/s between blocks. For configurations oriented at − 10° and − 20° towards the southwest (SW), wind velocities at the front blocks ranged from 3.4 to 4 m/s, gradually decreasing to 0.85 to 1.25 m/s at the rear, with inter-block velocities between 2 and 2.5 m/s. In contrast, orientations at + 30° and

Table 3 Boundary conditions for CFD simulation

Grid type	Grid spacing	Gridline merge tolerance (m)	Wind velocity (m/s)	Discretization scheme	Maximum number of outer iterations	Turbulence model	Direction (°)	Exposure	Site domain (m)
Non uniform	1.000	0.2000	3.5 m/s	Upwind	1000	2-k-e	260°	Semi urban	Length—1.5 Width—1.5 Height—1.5

Table 4 Human perception towards air velocity (m/s)

Air velocity at head level of the occupant (m/s)	Typical subjective occupant evaluation
0–0.05	Complain about stagnant air
0.05–0.25	Generally favourable (manufacturers of air outlet devices, e.g. base performance on 50 fpm air velocity in the occupied zone)
0.25–0.5	Awareness of air motion but still comfortable (e.g. some retail shops and stores) when the temperature of moving air is above room air temperature
0.5–1.02	Constance awareness of air motion but can be acceptable (e.g. some factories) if the air supply is intermittent and above air room temperature
> 1.02	Increasingly draft conditions with complaints about 'wind' disrupting a task, activity, etc.

Source Ghalam et al. (2021)

Table 5 Comfort index for wind velocity (m/s)

Wind velocity (m/s) at head level of occupant	Human sensation
0–0.05	Stagnant air
0.05–0.254	Favourable
0.254–0.51	Comfortable
0.51–1.02	Acceptable
1.02–1.5	Disrupting activities
> 1.5	Discomfort

Source Ghalam et al. (2021)

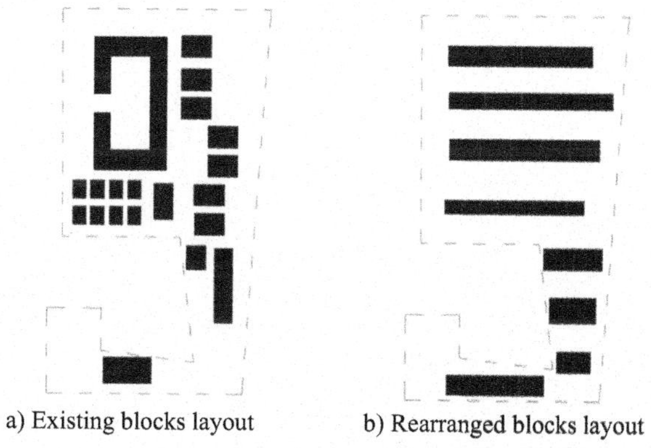

Fig. 3 **a** Existing blocks layout, **b** rearranged blocks layout

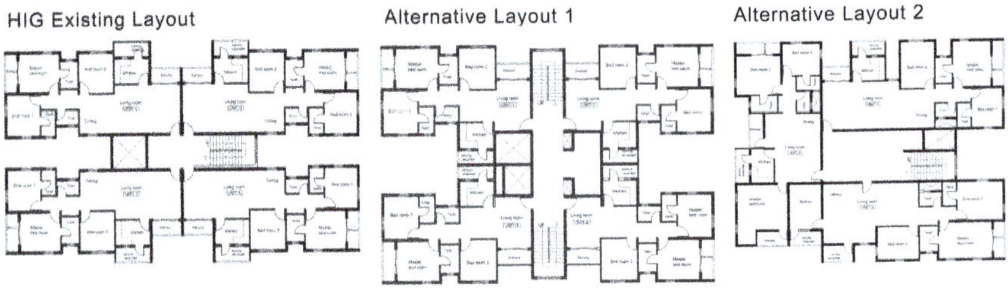

Fig. 4 Three interior layouts for analysing internal CFD

Table 6 CFD simulation for the Case A, B and C on the site with different orientation and block pattern

Building orientation	-20°	-10°	0°	30°	45°
Case A	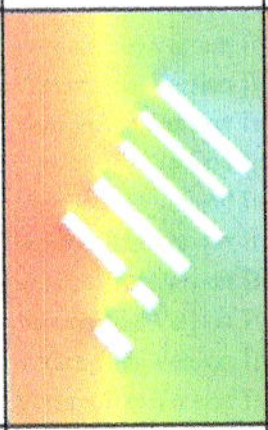		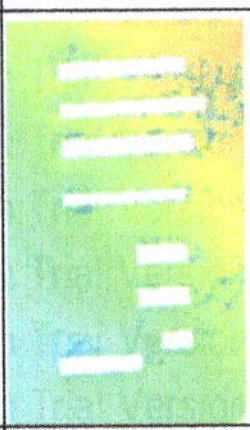		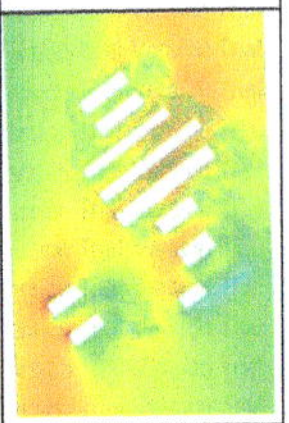
Distance between blocks	20m – 32m	20m – 32m	20m – 32m	20m – 32m	20m – 32m
Case B	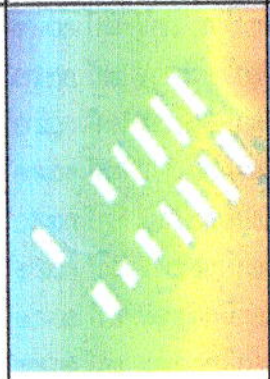				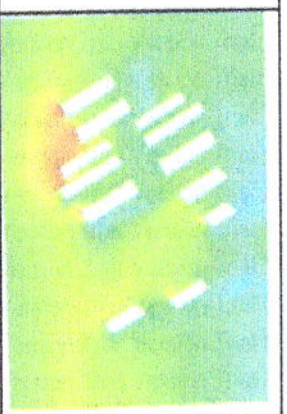
Case C	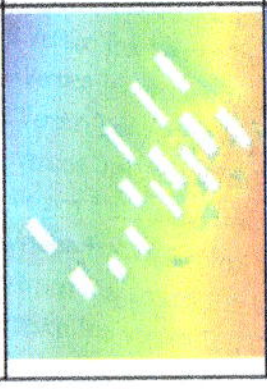	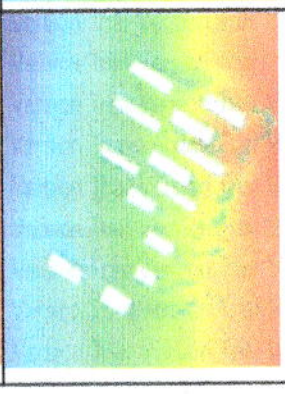	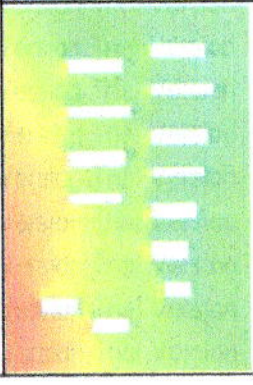	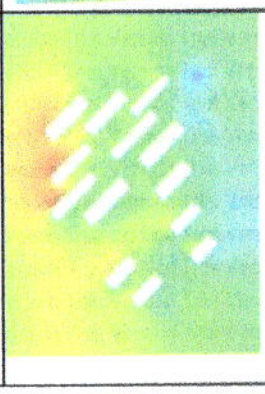	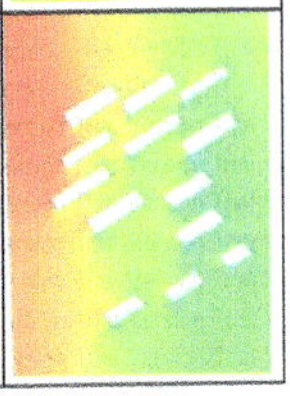

+ 45° towards the southeast (SE) maintained higher wind velocities at the front but dropped to 1.26 to 3 m/s at the rear.

Internal CFD simulations were carried out for the existing High-Income Group (HIG) block layout and two alternative designs, both oriented at + 45°, to assess indoor thermal comfort. Each HIG floor features four residential units, with the

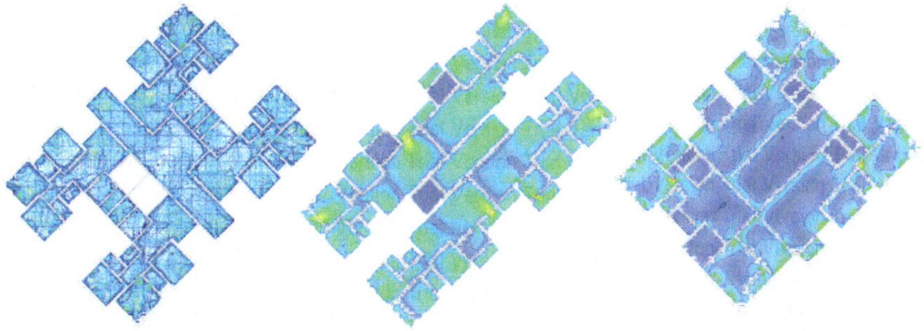

HIG existing layout
Size: 29.7m x 20.6m

b) Alternative Layout 1
Size: 33m x 18.2m

c) Alternative Layout 2
Size:26.6m x 21.3m

Fig. 5 CFD simulations for the internal layouts with + 45° orientation

alternatives maintaining the same overall area as the existing layout. Alternative 1 (Fig. 5) employs a linear arrangement of rooms, resulting in higher wind velocities of 0.05–1.75 m/s due to efficient space distribution. Alternative 2, designed with a more compact and squarer layout, performed less effectively.

This study focuses solely on airflow and thermal comfort analysis, excluding the impact of contextual buildings or interior furnishings. These findings provide valuable insights for optimising block arrangements and interior layouts to enhance natural ventilation and improve thermal comfort in residential complexes.

4 Conclusions

After studying the wind flow movement for the block arrangements and the response of internal layouts with the derived optimum orientation, the following summarizations are made:

1. The results indicate that the more suitable block orientation is between + 30° and + 45° tilted to southeast to capture prevailing southwest annual winds to increased velocity to ventilate the interior spaces. The long axis should be oriented between + 45° and + 60° from north axis.
2. The best block arrangement is staggered pattern arrangement so as to accentuate wind movement.
3. The arrangement of interior spaces in an important determinant of the indoor wind velocity, which has also been experimented in this paper concluding that the linear arrangement of interior spaces within a residential unit will receive the maximum wind velocity (0.05–1.75) m/s.

4. Both external and internal CFD are simulated without the consideration of the neighbourhood trees, buildings and interior furniture, which gives scope for further research. All the alternative interior layouts are provided with cross ventilation, where the windows facing the windward sides are smaller compared to the leeward facing windows so as to create a positive venturi effect within the interiors.

This paper provides insights on block arrangements with varying orientations and patterns, demonstrating how strategic configurations can harness natural wind flow for passive ventilation. These findings serve as a foundational guide for the preliminary design of residential complexes in hot and humid climates, such as Chennai.

References

Abbasi, S. (2021). Impact of wind-energy generation on climate. *Sustainable Energy, Grids and Networks.*

Asfour, O. S. (2010). Prediction of wind environment in different grouping patterns of housing blocks. *Energy and Buildings, 42*, 2061–2069.

BEE. (2020). *Energy Benchmarks for buildings*. The Bureau of Energy Efficiency, Government of India.

Daemei, A., Khalatbari, A., & Safari, H. (2016). Opening performance simulation in natural ventilation using design builder. *Energy Procedia,* 412–422.

Egan, M. D. (1975). *Concepts in thermal comfort*. Prentice Hall.

Esfahankalateh, A. T., et al. (2021). Achieving wind comfort through window design in residential buildings in cold climates, a case study in Tabriz city. *International Journal of Low-Carbon Technologies, 16*(2), 502–517.

Ghalam, N. Z., Farrokhzad, M., & Nazif, H. (2021). Investigation of optimal natural ventilation in residential complexes design for temperate and humid climates. *Sustainable Energy, Grids and Networks, 27*, 100500.

Hong, B., & Lin, B. (2015). Numerical studies of the outdoor wind environment an thermal comfort at pedestrian level in housing blocks with different building layout patterns and trees arrangement. *Reneewable Energy, 73*, 18–27.

Marija, E. T. (2007). Natural and mixed ventiation design via CFD and architectural modeling. *International Journal of Ventilation, 5*(4), 447–458.

Vijayalaxmi, J. (2009). The thermal comfort of a naturally ventilated house resulting from the evaporative cooling of a ceiling fan in the hot-humid climate of Chennai, India. *International Journal of Ventilation, 8*(2), 135–144.

Vijayalaxmi, J. (2013). Thermal comfort of a naturally ventilated house due to the evaporative cooling of ceiling fan in hot-humid climate of Chennai, India. *Journal of Architectural and Planning Research,* 1–22.

Yang, X., et al. (2019). Integrated Assessment of indoor and outdoor ventilation in street canyons with naturally buildings by various ventilation indexes. *Building Environment,* 106–528.

Zhou, C., Wang, Z., & Chen, Q. (2014). Design optimisation and field demonstration of natural ventilation for high-rise residential buildings. *Energy Build,* 475–476.

Comparative Analysis of the Baseline and Remodelled Scenario to Study the Presence of Green Infrastructure to Mitigate Heat Intensity: A Case of Vijayawada City

Prashanti Rao and Somaina Islary

Abstract The majority of India's tier-two cities are witnessing both vertical and horizontal expansion, regardless of their diverse topographical and climatic conditions. In addition to the expansion of cities, urbanization also plays a significant role, impacting the spatial and environmental conditions and contributing to environmental risks such as the Urban Heat Island (UHI) effect. Subsequently, multiple study findings have consistently shown that UHI effect is a result of an amalgamation of different crucial climatic factors present within the urban environment. Globally, various approaches are being tested to mitigate UHI effect. The green infrastructure (GI) approach is identified as one of the simplest, resilient, and most cost-effective approaches due to its easy applicability. Furthermore, this approach has been extensively examined in several studies conducted on a wider geographical scale, using tools like Geographic Information Systems (GIS), etc. This research intends to examine the impact of UHI for varied locations within a city, which have distinct topographic and physiographic conditions for extreme hot and humid climate. The study area is chosen as Vijayawada city, India; one of the fastest growing cities having distinct topography and physiographic conditions (hot and humid). This study was conducted at the microscale at three locations for mixed land use within the proximity of 2 km. Further, to simulate the base and remodelled scenarios of study areas, ENVI-met 5.5.1 lit version has been used. In this preview, data have been collected for dry bulb temperature, relative humidity, and wind speed on 11th November 2023 at 1:00 PM. This is simulated for five climatic parameters such as potential air temperature (PAT), mean radiant temperature (MRT), relative humidity (RH), wind speed (WS) and direct radiation short wave (DRSW). Comparative study showed that even within the closed proximity, all these five parameters are varying. Later green vegetation was introduced strategically and simulated to record its impact on heat intensity which eventually helps to understand the impact on UHI effect at microscale. This

P. Rao (✉) · S. Islary
School of Planning and Architecture Vijayawada, Architecture Department, Vijayawada, India
e-mail: prashanti.swe@spav.ac.in

S. Islary
e-mail: somainaislary@spav.ac.in

J. Vijayalaxmi (ed.), *Sustainable Building Performance*,
https://doi.org/10.1007/978-981-96-2428-7_10

paper intends to look into the critical understanding about resilience developed by its locational criteria to mitigate or to reduce the UHI effect at microlevel.

Keywords Baseline and remodelled scenario · Climatic factors · Green infrastructure · Resilience · Urban Heat Island (UHI) effect

1 Introduction

The Urban Heat Island (UHI) phenomenon poses substantial environmental hazards to those living in urban areas. The UHI effect is anticipated to be intensified by climate change. It is a distinct climatic phenomenon in which densely populated places exhibit higher temperatures compared to their surrounding regions. An important factor contributing to the development of the UHI phenomenon is the presence of extensive built-up surfaces, such as concrete and asphalt, which possess a significant heat capacity. The problem is being exacerbated by the use of materials with a low albedo. The procedure differs depending on the architectural characteristics and geographical circumstances of an urban area (Osei et al., 2022). In other words, the impact of UHI is a result of insufficient air movement, changes in the thermal properties of surface materials, and a decreased rate of evapotranspiration in metropolitan areas. These alterations cause an increased heat capacity, which leads to the storage of more energy and radiation, culminating in a subsequent increase in temperature. Furthermore, human activities lead to an augmented emission of heat in urban areas that is surpassing that of the natural surroundings (Marando et al., 2022). Introducing GI measures in this situation may somewhat reduce the intensity of UHI promoting a resilient urban environment and assisting in adapting to and mitigating climate change. In order to achieve this result, it is essential to systematically integrate GI into urban planning and policy. Nevertheless, this process depends on the existence of data that can be widely applied, easily accessible, and quantifiable. On the other hand, green regions with plants and soil that can hold water have the ability to absorb a substantial quantity of incoming radiation and then release it via evapotranspiration. Evaporation induces a cooling phenomenon, wherein the ensuing water vapour aids in reducing the ambient temperature. Studies have shown a correlation between the increase in urbanization and the average daily temperature rise. The arrangement of land cover attributes may significantly influence the extent of land surface temperature. Hence, the key component in reducing the effect of the UHI is the rise in the proportion of vegetation, which leads to a decrease in land surface temperature. Around the world, many strategies are being experimented with to alleviate the UHI phenomenon. The GI approach is recognized as one of the most simple, resilient, and cost-effective methods owing to its simple implementation (Osei et al., 2022). Moreover, this methodology has undergone thorough scrutiny in several research done on a broader geographic scope, using techniques such as GIS. To get a thorough understanding of the severity of urban heat and effective approaches to mitigate high temperatures, it is important to evaluate a neighbourhood model that assesses the ecosystem service of microclimate management.

A low cost and non-time-consuming approach to estimate UHI intensity over space and time is urgently needed for researchers as well as policy-makers. Trees have the most significant influence on cooling via green infrastructure, making them the predominant approach for implementing green infrastructure. The majority of research has also shown a consistent correlation between the ability of trees to reduce temperature and improve human thermal comfort. Trees emit water vapour into the atmosphere via evapotranspiration, resulting in a decrease in temperature and an enhancement in thermal comfort. In order to decrease the temperature by 1 °C, it is recommended that trees occupy a minimum of one-third of the total area (Alves et al., 2022). Similarly, the capacity of a waterbody or watercourse to alter nearby temperatures is influenced by its intrinsic characteristics as well as its interactions with the surrounding climatic conditions. The high thermal inertia of water, due to its enthalpy of vapourization, plays a crucial role in regulating temperatures and temporal fluctuations. This allows bodies of water to function as effective thermal buffers.

Hence, this research intends to examine the impact of UHI for distinct topography, and hot and humid climatic conditions for mixed uses-built forms. The model replicates the disparity in temperature, humidity, and wind speed between a standard (Baseline) scenario when compared with simulated (remodelled) scenario having more vegetation and projecting the impact of GI in reducing UHI effects in various metropolitan settings.

2 Study Method

2.1 *Study Area*

The city of Vijayawada (as shown in Fig. 1) is located at the heart of Andhra Pradesh in India. It is a historically significant city that lies on the banks of the Krishna River. Its geographical coordinates are 16.0311° N latitude and 80.391° E longitude. This city has been chosen because of its distinct topography; most of it is flat land surrounded by few small to medium-sized hills within the city core. The city has a tropical climate with hot summers and moderate winters. The temperature can soar to a scorching 47 °C during May–June, while the winter temperature ranges from 20 to 27 °C. Both these distinct factors, topography and hot-humid climate, develop a kind of heat canyon throughout the year, thus impacting the microclimate of the place. The average humidity stands at 78%, and the city receives an annual rainfall of 103 cm from both the southwest and northeast monsoons. It is a vital railway junction, connecting all the states in the country. For the study purpose, Indrakeeladri Hill, and its three microscale spatial units (as shown in Figs. 1, 2, and 3a–c) have been chosen to measure the impact of GI on UHI, based on the situations

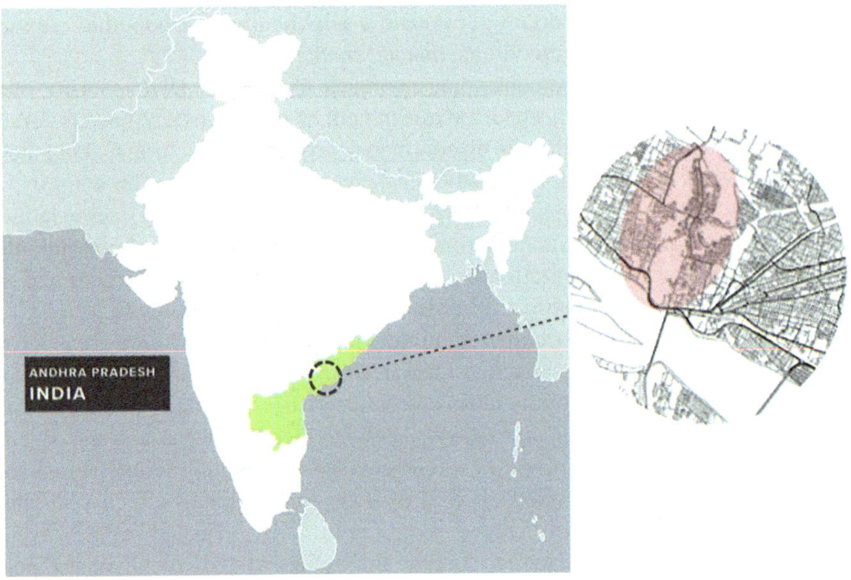

Fig. 1 Location of site

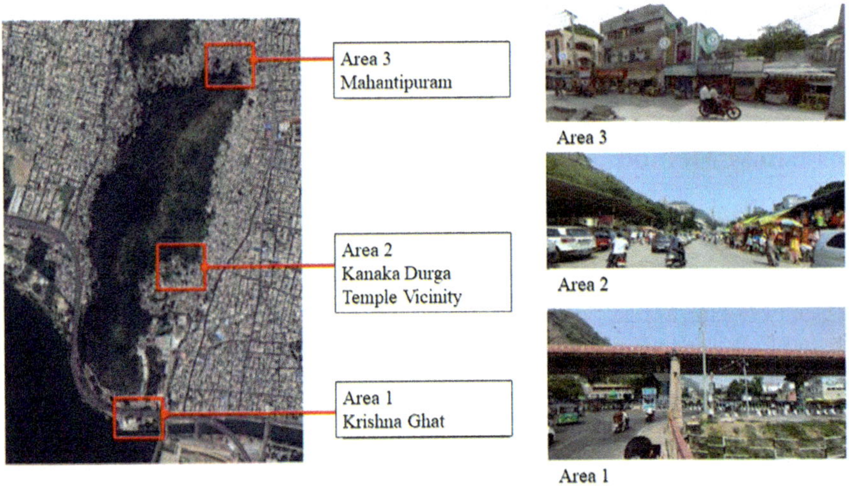

Fig. 2 Location of three study areas for simulation

around the selected built environments. The area around the Kanaka Durga Temple is located on Indrakeeladri Hill, a small hill in the city of Vijayawada. The hill is situated on the banks of the Krishna River. The hill is steep and rocky and has a height of around 180 m. It is surrounded by lush greenery and offers a breath-taking view of the surrounding landscape, including the Prakasam Barrage, a major dam on the Krishna River. The area surrounding the Temple and the Prakasam Barrage has a hot and humid climate and a mix of scrub and dry deciduous forest vegetation. The area is also characterized by the presence of igneous and sedimentary rocks and is prone to earthquakes. On the other side of the hill, there are residential settlements that have settled near the foot hills.

2.1.1 Area 1

Krishna Ghat, Vijayawada—Durga Agraharam, located along the Krishna Ghat and adjacent to the Prakasam Barrage is a historic area located in the heart of Vijayawada. The area is characterized by narrow streets, old buildings, and a mix of commercial and residential land uses.

2.1.2 Area 2

Kanaka Durga Temple Vicinity, Vijayawada—Kanaka Durga Temple is a famous Hindu temple located on the banks of the Krishna River in Vijayawada. The area around the temple is a major tourist attraction and is characterized by a mix of commercial and religious land uses.

2.1.3 Area 3

Mahantipuram, Vijayawada—It is a relatively new residential area located in the northern part of Vijayawada. The area is characterized by wide streets, low-rise buildings, and a mix of residential and open spaces.

2.2 Methodology

To understand the heat intensity of the three selected areas and it impact on UHI, ENVI-met 5.5.1 modelling tool has been used, to simulate the baseline and remodelled scenarios. In this preview, as a first step, field data have been collected for Dry Bulb Temperature, Relative Humidity, and Wind Speed on 11th November 2023 at 1:00 PM as shown in Table 1 for analysing accuracy for simulated data. Secondly, simulation has been carried out with the input parameters as shown in Fig. 4 for all three areas, which in turn resulted in output for five climatic parameters such as potential air temperature (PAT), mean radiant temperature (MRT), relative humidity (RH), wind speed (WS), and direct radiation short wave (DRSW) to understand the heat intensity of each location for 6 h. Finally, the same input file has been simulated with strategic introduction of one GI component, i.e., additional vegetation in the existing vacant land, vertical greenery, and rooftop green (as shown in Table 2) to

(a)

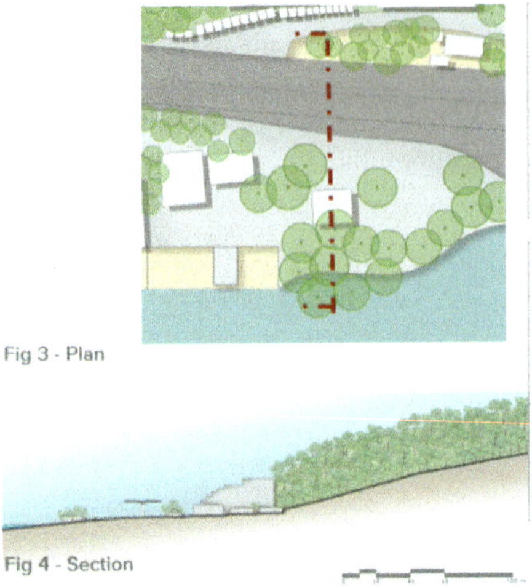

(b)

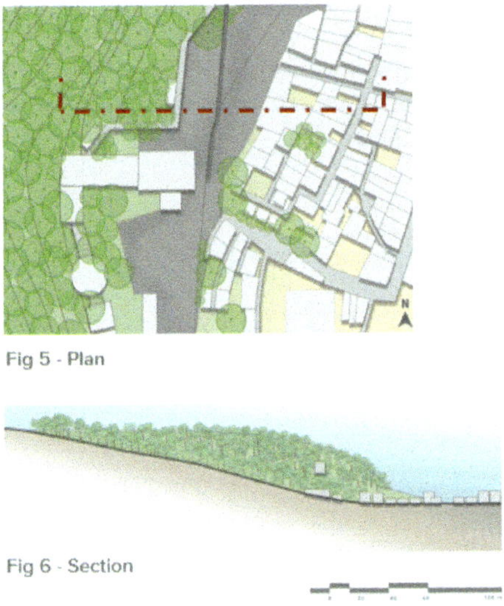

Fig. 3 **a** Baseline scenario, Area 1(plan and section)—Krishna Ghat, Vijayawada, **b** baseline scenario, Area 2 (plan and section)—Kanaka Durga Temple Vicinity, Vijayawada, **c** baseline scenario, Area 3 (plan and section)—Mahantipuram, Vijayawada

(c)

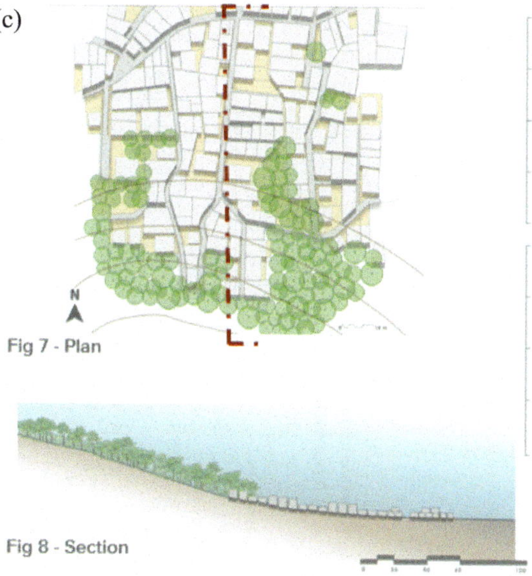

Fig. 3 (continued)

record its impact on heat intensity which eventually helps to understand the impact on UHI effect at microscale as a major component of GI.

2.3 *Limitations*

In this study, ENVI-met still has some constraints, particularly regarding the thermal performance of building walls. This restriction may impact the microclimate since the model assumes a single material, which deviates from real-world conditions. Furthermore, the model adjusts the simulations of air temperature and relative humidity, while wind conditions and cloudiness stay relatively constant. Another constraint arises in connection to MRT and the assessment of radiative fluxes, which exhibits a minor degree of inaccuracy, despite the presence of a robust correlation between the simulated and observed data. Besides, the model does not take into consideration anthropogenic heating, automobiles, and mechanical cooling systems, which have the potential to modify temperature conditions.

Table 1 Microstudy area details and field measurements on 11th November 2023 at 1:00 PM

S. No.	Study area	Area in m^2	Type of built use	% Available green	Proximity to blue in m	Relative humidity (%)-hygrometer	Temp (DBT)-Degree Celsius—Testo Metre 850	Wind speed (m/s) @ 1.5 m (Anemometer)
1	Area 1	15,625	Mixed use	34%	0.00	53%	32.1 °C	1.9–2.3 m/s
2	Area 2	15.625	Mixed use	42%	650.00	50%	32.1 °C	1.9–2.3 m/s
3	Area 3	15,625	Mixed use	18%	1600.00	59%	28 °C	0.3–0.5 m/s

Source Author

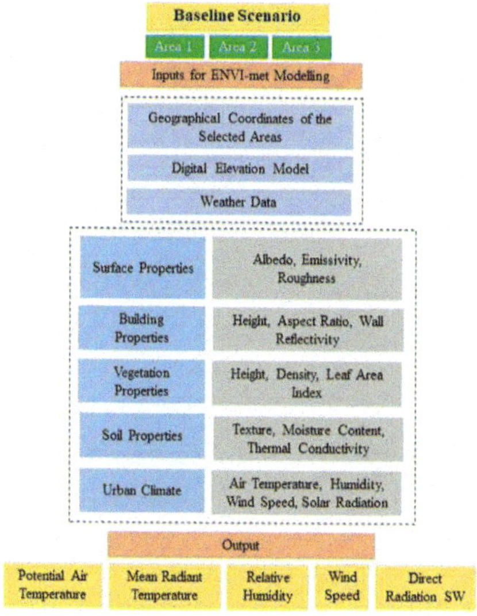

Fig. 4 Methodology

Table 2 Comparative values of percentage of green for both scenarios

Study area	Name of location	Baseline scenario	RRemodelled Scenario		
		Existing green on ground (in %)	Added green on vacant land (in %)	Added green Façade (in %)	Added rooftop green (in %)
Area 1	Krishna Ghat, Vijayawada	22	8	2	1
Area 2	Kanaka Durga Temple Vicinity, Vijayawada	38	0.4	0.6	1
Area 3	Mahanti Puram, Vijayawada	18	2	0.6	1.4

Source Author

3 Analysis and Results

3.1 Baseline and Remodelled Simulation Scenarios

As shown in Fig. 5, all three study areas are provided as input data along with the essential climate, location and grid parameters for simulation process, dated 11th November 2023. This simulation process was conducted for six hours as area for

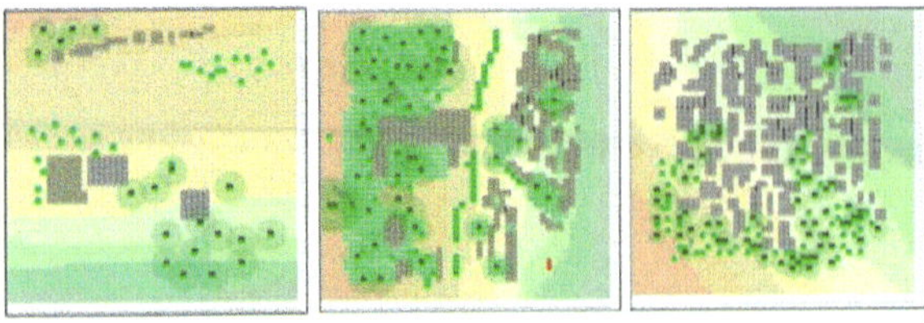

Baseline simulation model input for Area-1 ,2 and 3

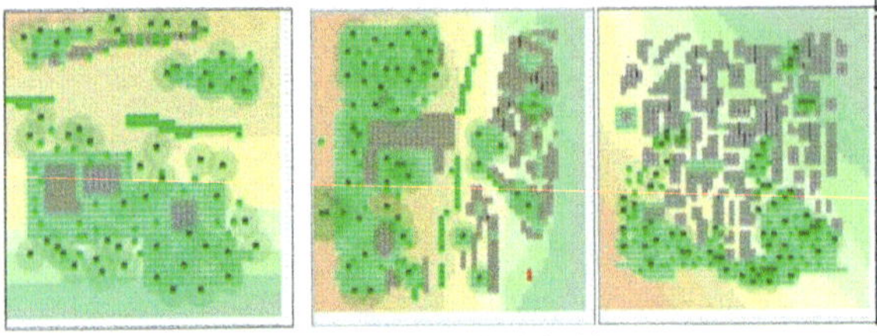

Remodelled simulation model input for Area-1 ,2 and 3

Type	Parameter	Value
Air Temperature	Initial Temperature atmosphere (K)	300 (26.8 °)
Wind	Wind Direction in 10 m ab. Ground (m/s)	2.24
Humidity	Relative Humidity in 2m(%)	78
Location	Location	Vijayawada
	Latitude	Lat-16.51 deg N, S
	Longitude	Long -80.61 deg W, E

Study Area	Location	Grid Dimension	Model Dimension
1	Prakassam Barrage, Vijayawada, A.P	2X2X3	50x50x25
2	Kanaka Durga Temple, Vijayawada	2X2X3	50x50x25
3	MahantiPuram, Vijayawada	2X2X3	50x50x25

Fig. 5 Baseline and remodelled scenarios simulation modelling—input and output

input is at neighbourhood level to obtain the simulation results for all five parameters, i.e., PAT, MRT, RH, WS, and DRSW. Obtained results show little variations amongst each other when compared as shown in Fig. 6a–e as well as in Table 3.

3.1.1 Mean Radiant Temperature (MRT)

MRT is an exchange of radiant heat between humans and their surrounding environment with a view to understand the influence of surface temperature on personal comfort. Mean radiant temperature is a parameter affected by surfaces' albedo and by the shadowing from both buildings and trees. The behaviour of the system varies during the day due to the impact of solar radiation load. Furthermore, ENVI-met may not accurately represent the intricate characteristics of surface albedos and plant attributes (such as leaf area index), which might explain the disparities seen between the baseline and modelled outcomes. Hence, in the baseline scenario simulation, MRT is more for the study Area 1 when compared with other two respective study areas. Whereas in remodelled scenario, due to introduction of green vegetation as per the availability of vacant land and built surfaces MRT becomes comparatively lower for Area 1, not much change for Area 2 and comparatively more in Area 3 as shown in Fig. 6a. In Area 3, maximum temperature is more around the built units and minimum near the already existing green and proposed green, whereas in Area 1 the GI intervention enhanced the cooling impact more in wider area and reduced MRT, thus reduced the heat intensity and developed resilience as a mitigation method to reduce the impact of UHI.

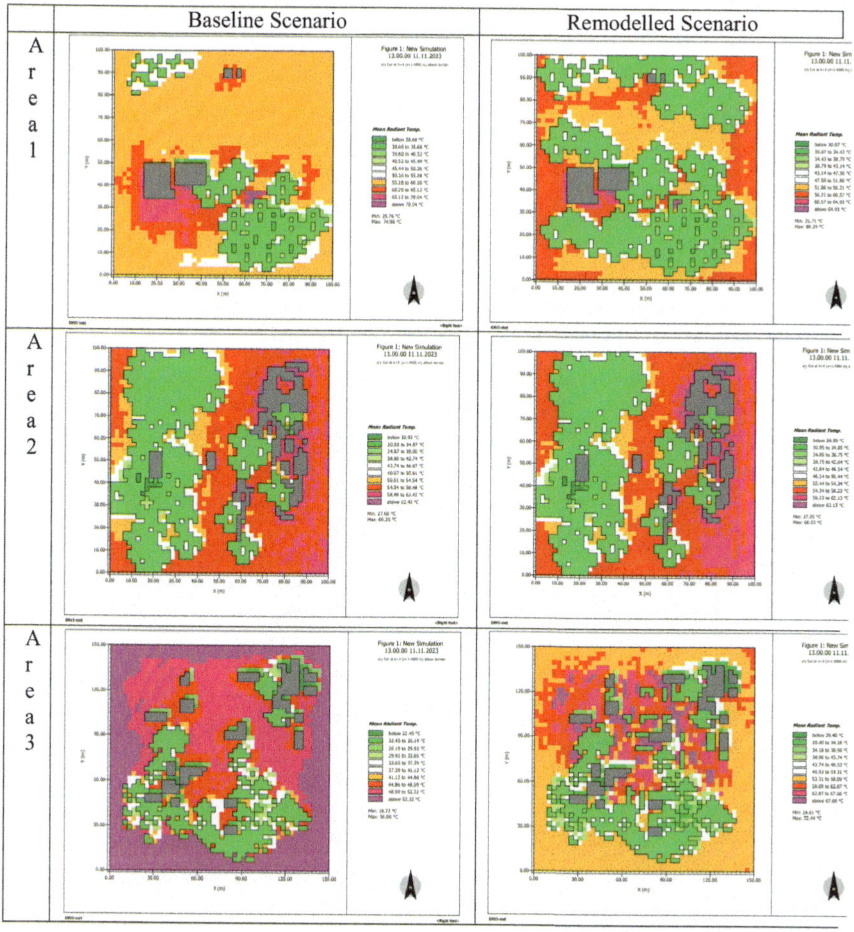

(a)

Fig. 6 **a** Baseline and remodelled scenarios models for mean radiant temperature for all, **b** baseline and remodelled scenarios models for potential air temperature for all three locations, **c** baseline and simulated model for relative humidity for all three locations, **d** baseline and remodelled scenarios models for wind speed for all three locations, **e** baseline and remodelled scenarios models for direct radiation SW for all three locations

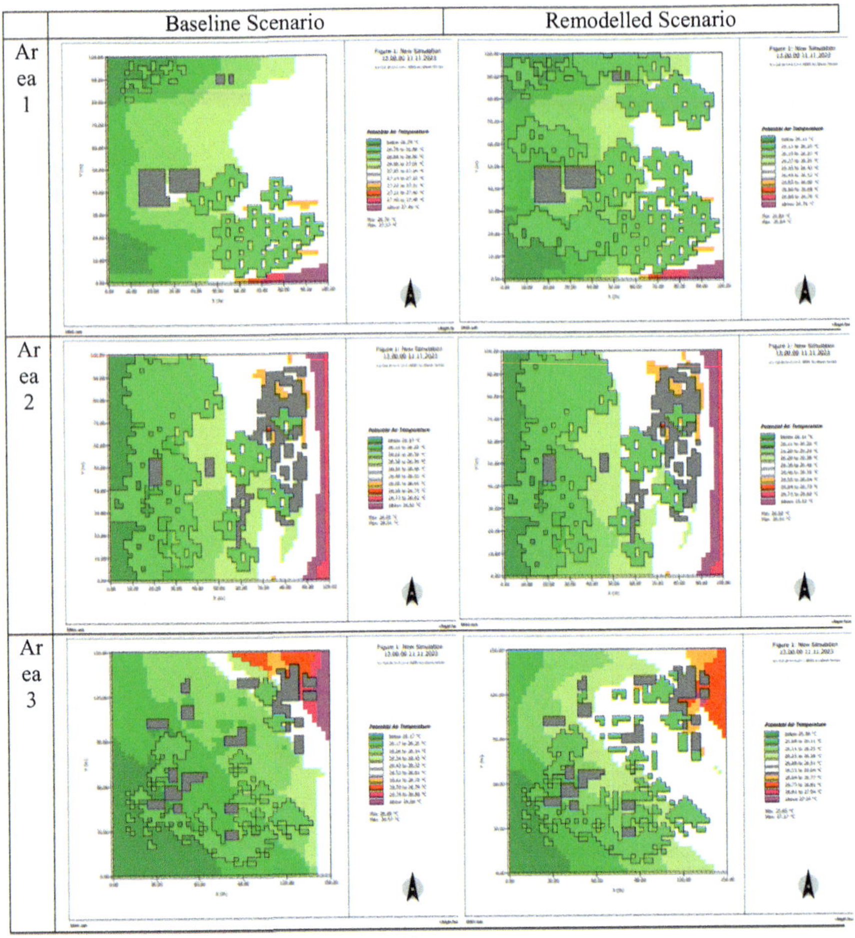

(b)

Fig. 6 (continued)

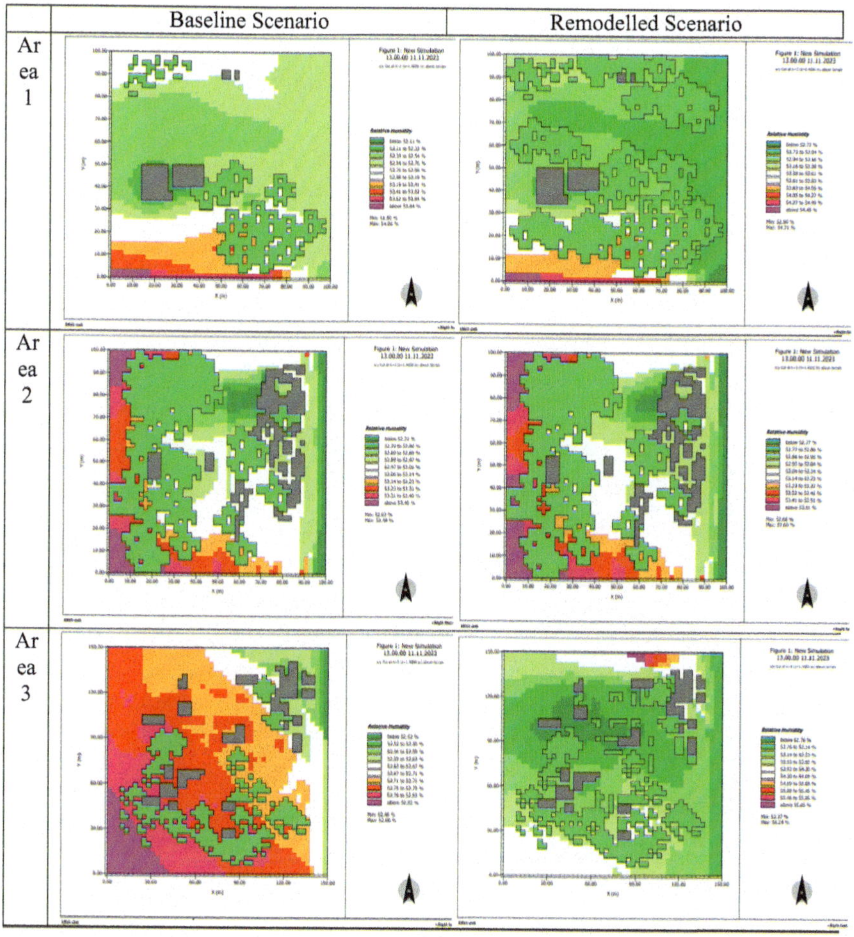

(c)

Fig. 6 (continued)

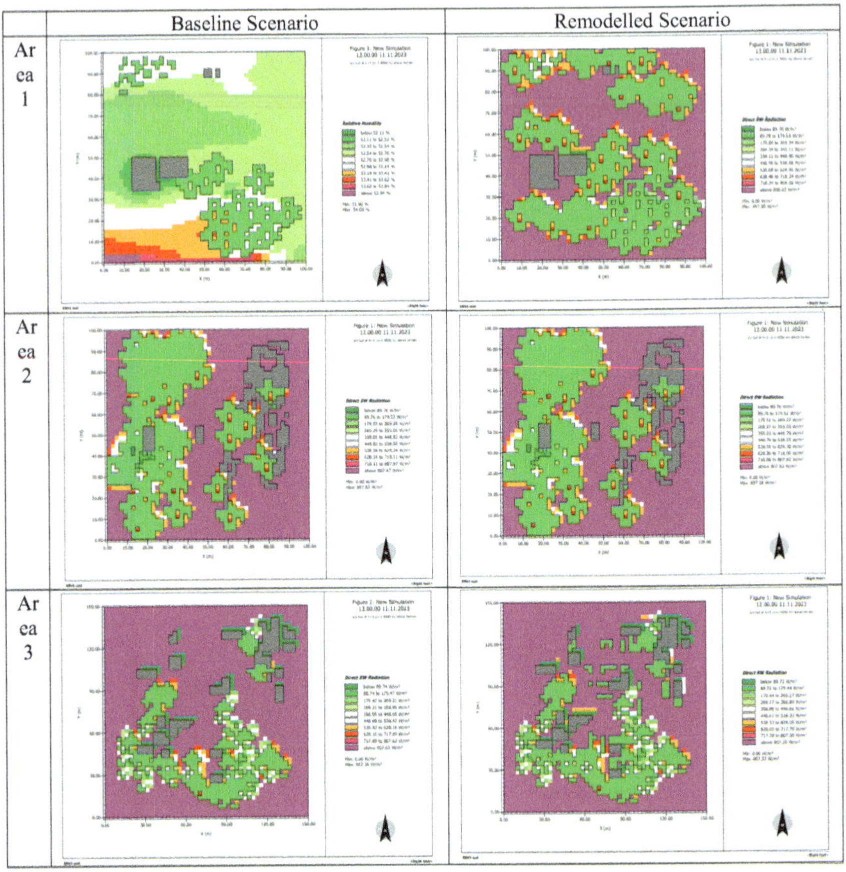

(d)

Fig. 6 (continued)

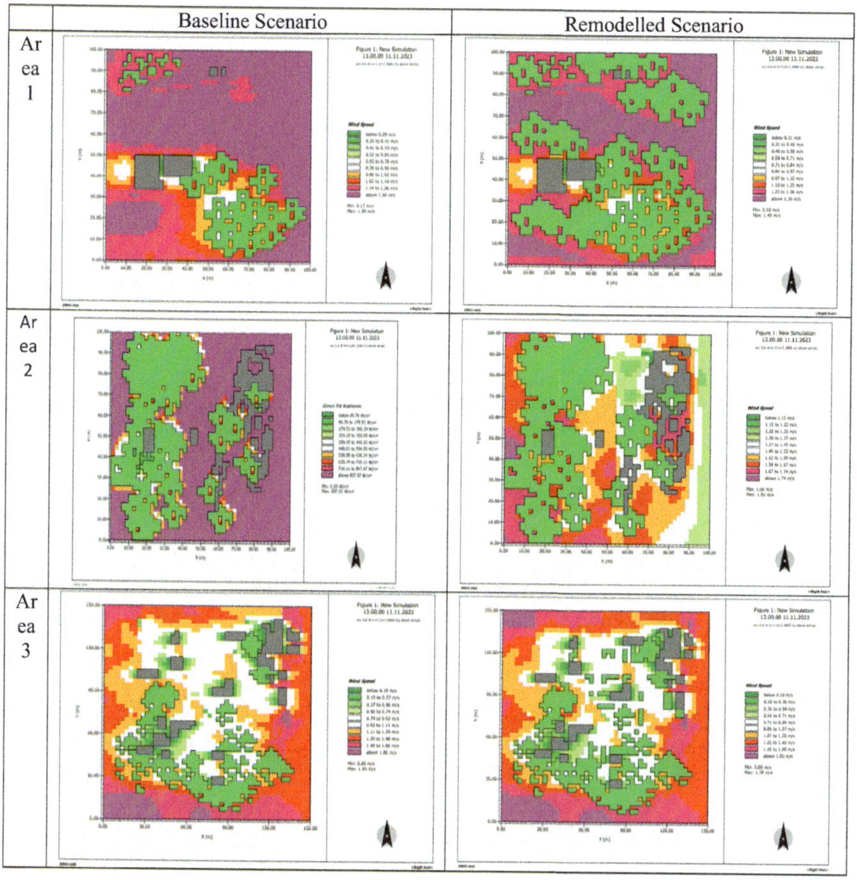

(e)

Fig. 6 (continued)

3.1.2 Potential Air Temperature (PAT)

The maps indicate that the greatest air temperatures were projected to occur over asphalt and walkways, while the lowest air temperatures were simulated on grass surfaces, under trees, and in other shaded areas provided by both buildings and trees. At 01:00 PM, the simulated temperatures reach their peak, ranging from 25.85 °C to 26.03 °C, with the largest concentrations seen in the traffic circle and asphalted area. Temperature rose, particularly near surfaces with low albedo, reaching peak temperatures of 26.84 °C and 27.57 °C for the base and remodelled scenarios, respectively. As per the simulation results, the PAT is more for Area 1 when compared to other two study areas. But according to the remodelled scenario, the situation is reversed, where PAT is decreased by 1 °C. Area 3 and Area 2 have shown very negligible change whereas study Area 1 has shown drop by 1 °C. The simulated scenarios

Table 3 Comparative values for all five climatic parameters for both scenarios

Study area	Name of location	Potential air temp for base scenario		Potential air temp for remodelled scenario	
		Highest value in (°C)	Lowest value in (°C)	Highest value in (°C)	Lowest value in (°C)
Area 1	Krishna Ghat, Vijayawada	27.57	26.7	26.84	26.03
Area 2	Kanaka Durga Temple Vicinity, Vijayawada	26.91	26.02	26.91	26.05
Area 3	Mahantipuram, Vijayawada	26.97	26.08	27.57	25.85
Study area	Name of location	Mean radiant temp for base scenario		Mean Radiant Temperature for Remodelled Scenario	
		Highest value in (°C)	Lowest value in (°C)	Highest value in (°C)	Lowest value in (°C)
Area 1	Krishna Ghat, Vijayawada	74.96	25.76	69.29	25.71
Area 2	Kanaka Durga Temple Vicinity, Vijayawada	66.03	27.05	66.35	27
Area 3	Mahantipuram, Vijayawada	56.06	18.72	72.44	24.6
Study area	Name of location	Relative humidity for base scenario		Relative humidity for remodelled scenario	
		Highest value in %	Lowest value in %	Highest value in %	Lowest value in %
Area 1	Krishna Ghat, Vijayawada	54.06	51.90	54.71	52.50
Area 2	Kanaka Durga Temple Vicinity, Vijayawada	53.60	52.68	53.48	52.63
Area 3	Mahantipuram, Vijayawada	52.86	52.48	56.24	52.37
Study area	Name of location	Wind speed for base scenario		Wind speed for remodelled scenario	
		Highest value in m/s	Lowest value in m/s	Highest value in m/s	Lowest value in m/s
Area 1	Krishna Ghat, Vijayawada	1.38	0.17	1.49	0.18
Area 2	Kanaka Durga Temple Vicinity, Vijayawada	1.82	1.08	1.81	1.10
Area 3	Mahantipuram, Vijayawada	1.85	0	1.78	0

(continued)

Table 3 (continued)

Study area	Name of location	Solar radiation SW for base scenario		Solar radiation SW for remodelled scenario	
		Highest value in watt/m^2	Lowest value in watt/m^2	Highest value in watt/m^2	Lowest value in watt/m^2
Area 1	Krishna Ghat, Vijayawada	894.70	0	897.80	0
Area 2	Kanaka Durga Temple Vicinity, Vijayawada	897.58	0	897.63	0
Area 3	Mahantipuram, Vijayawada	897.36	0	897.22	0

show that the inclusion of green spaces and GI interventions can potentially reduce air temperature, particularly during the early afternoon when solar radiation is more intense and surfaces have been warming up throughout the preceding hours. In simulations conducted in tropical climates (Morakinyo and Lam.) obtained comparable outcomes, emphasizing the significance of tree selection as distinct characterization parameters along with leaf area index values, leaf density distribution by height, and the planting pattern or arrangement, all of which can influence the potential benefits of trees.

3.1.3 Relative Humidity (RH)

The diurnal variation in RH can be primarily attributed to its correlation with air temperature and pressure. As the temperature drops, the air's capacity to hold water vapour decreases, leading to an increase in relative humidity. As per the simulation results, the RH is more for Area 1 when compared to other two study areas. But according to the remodelled scenario, the situation has not changed much for Areas 1 and 2, whereas the RH percentage has increased in Area 3.

3.1.4 Wind Speed (WS)

As per the simulation results, the WS is more for Area 3 when compared to other two study areas. According to the remodelled scenario, all three study areas have shown an increase in wind speed. The behaviour of air temperature may be significantly influenced by wind speed, since it can impact dynamic processes like convection. Unlike the other variables, ENVI-met produces constant values for wind speed throughout the whole simulation, therefore failing to capture fluctuations in wind speed during the day.

3.1.5 Direct Radiation Short Wave (DRSW)

Direct solar radiation refers to solar radiation received in the form of direct solar beams not blocked by clouds or other atmospheric obstacles. As per the existing simulation results, the DRSW is less for Area 1 when compared to other two study areas. According to the remodelled scenario, all three study areas have shown negligible change.

3.2 Impact of Topography and Physiography on Study Areas for Remodelled Scenario

When the remodelled scenario was compared to the existing scenario, it was found that all the three study areas reacted to the simulated results in a different way. Since Vijayawada is not located near a mountain range or a large body of water, the terms windward and leeward are not commonly used to describe the city's geography. But if the location of study area is considered, it is in the proximity of Prakasam Barrage, Krishna River, and Indrakeeladri hill. This creates a microclimate of windward and leeward situation. The study Area 1 is on windward side, study Area 2 in the cusp of windward-leeward side, and study Area 3 at leeward side. In remodelled scenario, more vacant land was available for study Area 1 for GI application when compared to study Area 3. Since study Area 2 has minimal vacant space for GI application on land parcel, vertical greening and terrace garden interventions are introduced for remodelling purpose. The impact of locational criteria was recorded in simulated results for MRT and PAT. In both the cases after increasing the percentage of GI intervention, it has an shown increase in the values for study Area 3 for remodelled scenario.

In baseline scenario for Area 1, despite having more vacant land, the presence of sealed surfaces in terms of tar roads and concrete flyovers accumulated the heat that had been radiating at 1:00 PM. Hence because of this existing scenario, simulation for MRT and PAT recorded more for the study Area 1 when compared with other two respective study areas. When Area 1 scenario was remodelled, trees were introduced, which resulted in an expansion of the cool zones and the MRT and PAT were lowered. These disparities may be explained by the fact that, trees partially deflected the sun's beams, resulting in reduced solar incidence, the same was recorded for DRSW. It has been established already through previous studies that the presence of plants in the built environment may effectively regulate the temperature of the air. In similar manner, due to its windward side location, the prevailing cool air throughout the day, in presence of trees as per remodelled simulation recorded increase in WS. Further, reduction in RH value was recorded compared to other two study scenarios, despite its relative closeness to water body, as trees exhale water vapour through a process called evapotranspiration, which helps to bring down the temperature of the air. This, in turn, helps to preserve the ability of the vegetation to contribute to the thermal comfort of the surrounding area.

In case of Area 2 for existing as well as for remodelled scenario simulations, it has been recorded that the values were almost similar for both, which is because of two important reasons. Firstly, very less vacant land was available for GI interventions between built environment and thus intervention was done for vertical greening and roof top green. But these interventions have shown less effectivity when compared with tree plantation and ground cover as part of GI. Secondly, due to the crossover of windward and leeward situation, impact of cold and dry wind nullified its impact for all five climatic parameters (MRT, PAT, RH, WS and DRSW) for which it is simulated. Hence, less resilience has been shown by Area 2 towards mitigation for UHI effect, due to no change in heat intensity, as less plantation opportunity, increased evapotranspiration, which contributed to increases in RH.

Area 3 in this study plays a very important role to understand the role of GI interventions for enhancing resilience to mitigate the UHI effect. This is in leeward side of Indrakeeladri hill, hence prone to dry air. In addition to that, due to varied land use activities like residential, commercial and transportation, and impervious surfaces, GI interventions alone is not sufficient to enhance resilience to mitigate the UHI effect. Figure 3c plan and section of study area clearly depicts less availability of the vacant land for plantation. Even if it is planted, evapotranspiration is more predominant than transpiration due to more heat radiation from built surfaces. Due to ENVI-met's inability to account for significant fluctuations in wind speed and direction inputs during the simulation, the simulated wind speeds seldom fluctuated throughout the day. Near the buildings, wind speeds were consistently lower throughout all time periods, but in open regions exposed to the main wind, speeds were consistently greater. The results suggest that the addition of vegetation in the remodelled scenario did not induce considerable changes. This result can be justified because the model was simulated by a single input value of wind speed and direction, southeast of the maps, without providing visible changes throughout the day.

4 Conclusion

This paper investigated the critical understanding about resilience developed by its locational criteria like topography and physiography for trapping or generating heat by comparing the simulated results for five climatic parameters by linking GI approach to mitigate or to reduce the UHI effect at microlevel. This research used climate modelling techniques to assess the possible advantages of integrating GI initiatives inside the core densely populated region of Vijayawada, using ENVI-met of Lite Version for simulation. This modelling programme enabled the estimation of distinct patterns in the behaviour of PAT, MRT, RH, WS, and DRSW for two scenarios: a baseline scenario and a GI remodelled scenario for three locations. When comparing these two scenarios, the findings indicate that the development of GI led to mixed results during the midday of 11th November 2023.

Further, for microlevel studies apart from climatic factors, physiographic conditions also play an important role in developing strategies for GI intervention. It is evident from this research that three study areas responded differently towards GI interventions as shown below:

Area 1—Positive impact of GI intervention has been observed because of windward side of Indrakeeladri hills, availability of more vacant place (no major built uses) and proximity to the water body as well as available green patches.
Area 2—Neutral and negligible change of GI intervention has been observed due to cusp of windward and leeward side of Indrakeeladri hills, availability of less vacant place and mixed built use. Apart from that, distance from waterbody and less percentage of available green patches are also among the contributing factors.
Area 3—Both positive and negative impact of GI intervention have been observed due to leeward side of Indrakeeladri hills, availability of less vacant place and mixed built use. The absence of waterbody as well as less percentage of available green space have also contributed to this impact.

These results indicate that for microlevel GI interventions, physiographic conditions may also be considered. For that location, instead of providing centralized solutions, decentralized solutions might be more effective.

The research found that GI interventions were successful in transforming cities into more environmentally friendly and creative models. This approach has considerable promise in mitigating the UHI impact in urban settings, hence enhancing the resilience and liveability of cities.

Acknowledgements The authors would like to thank Ar. Nischita, alumni of School of Planning and Architecture Vijayawada (SPAV) for assisting in documentation.

References

Alves, F. M., Gonçalves, A., & Rosario del Caz-Enjuto, M. (2022). The use of envi-met for the assessment of nature-based solutions' potential benefits in industrial parks—A case study of Argales industrial park (Valladolid, Spain). *Infrastructures, 7*(6), 85. https://doi.org/10.3390/infrastructures7060085

Marando, F., Heris, M. P., Zulian, G., Udías, A., Mentaschi, L., Chrysoulakis, N., Parastatidis, D., & Maes, J. (2022). Urban heat island mitigation by green infrastructure in European functional urban areas. *Sustainable Cities and Society, 77*, 103564. https://doi.org/10.1016/j.scs.2021.103564

Osei, G., Pascale, F., Delle-Odeleye, N., & Pooley, A. (2022). Green infrastructure. In *The Palgrave encyclopedia of urban and regional futures* (pp. 737–746). https://doi.org/10.1007/978-3-030-87745-3_260

The manufacturer's authorised representative in the EU is Springer Nature Customer Service Centre GmbH, Europaplatz 3, 69115 Heidelberg, Germany. If you have any concerns regarding our products, please contact ProductSafety@springernature.com

Printed and bound by CPI Group (UK) Ltd, Croydon, CR0 4YY
13/10/2026
02232626-0001